POLYPROPYLENE
FIBERS AND FILMS

Anthony V. Galanti, M.S.
and
Charles L. Mantell, Ph.D.
Newark College of Engineering
Newark, New Jersey

Springer Science+Business Media, LLC
1965

ISBN 978-1-4899-2824-5   ISBN 978-1-4899-2822-1 (eBook)
DOI 10.1007/978-1-4899-2822-1

Library of Congress Catalog Card Number 65-26813

©1965 Springer Science+Business Media New York
Originally published by Plenum Press in 1965.

Preface

This book represents a compilation and correlation of pertinent information currently available on polypropylene fibers and films. Specifically, the information presented considers the effects of fiber and film processing conditions upon polypropylene fiber properties as well as the engineering properties of polypropylene relative to other commercial fibers. The data on polypropylene fibers were obtained almost entirely from recent technical periodicals, reports, and technical literature of various polypropylene manufacturers. Since much of the original work on polypropylene was conducted by the Montecatini Company in Italy, several pertinent trade journals were foreign-based and required translation. Reference is made to sources of information indicated in the appended list of references for the tables and figures; many figures were reproduced as they appeared in the original articles.

When available, the origin of a fiber used in a specific analysis is presented by indicating the manufacturer's trademark; a list of these trademarks as well as several definitions are given in the appended section "Definitions and Fiber Trademarks."

The more general term "fiber" is used in the work to include both monofilaments which are relatively coarse fibers approximately 40–1000 den, and textile fibers which have a denier between 1–15. Multifilament yarns consist of a group of textile fibers assembled together to form a single thread. Since fibers are widely utilized in the form of yarns, considerable information on the properties of various yarns relative to polypropylene yarn is presented.

iii

The information contained in this work on polypropylene fibers as developed by various manufacturers is gratefully acknowledged. The excellent publications on the development work on polypropylene fibers and films by the Montecatini Company and the recent study by the Southern Research Institute are especially noteworthy.

Introduction

One of the latest members of the rapidly growing thermo-plastic polymer family which appears capable of successfully competing with the currently saturated textile and chemical markets is polypropylene fiber. The objectives of this book are to (1) examine the effects of polymer characteristics and fiber processing conditions on the properties of polypropylene fibers and (2) correlate all existing information on the physical, chemical, and mechanical properties of polypropylene fibers and comparatively evaluate this material with other fibers, natural and synthetic.

Polypropylene is the first member of a new group of polymers prepared by a mechanism defined as "stereospecific" polymerization. From a simple monomer, this technique produces a polypropylene with an exceptionally uniform molecular structure which imparts outstanding engineering properties into the polymer. Such structure regularity can be varied to tailor the properties of the polymer to best satisfy a given requirement.

The low costs of propylene monomer and the polymerization process give propylene a cost advantage over similar products. In addition, polypropylene fibers, because of their structural uniqueness, exhibit outstanding physical properties relative to other commercial fibers. The density of polypropylene is the lowest of any fiber available; supertenacity polypropylene fibers have been prepared that exceed the strength of all commercial fibers – including the much more expensive nylons. Polypropylene fibers also excel in other important physical properties, such as toughness, resilience, permeability, chemical resistance, and abrasion resistance.

The major problem with regard to widespread use of polypropylene is limited dyeability, a characteristic which stems
from the inherent inertness of the polypropylene structure to
permeants. However, in view of the major research effort de-
voted to this problem, it is reasonable to expect that an
answer is forthcoming.

In summary, the relatively low polymer cost and outstanding
properties of polypropylene fibers rank this material as one
of the important fibers of the future.

Contents

I. Polymerization of Propylene

A. DEVELOPMENT OF POLYPROPYLENE

For a long time, the polymers of propylene, an inexpensive petroleum derivative, were known only as viscous oils of little commercial usefulness. These oils, With a viscosity range dependent upon the molecular weight of the polymer, were not crystallized by the methods of polymerization known to produce a crystalline structure in such materials as ethylene, vinylidene chloride, and perfluoroethylene. However, in 1954, Professor G. Natta discovered a new polymerization mechanism which could transform the random structural arrangement of these "noncrystallizable" polymers into structures of high chemical and geometrical regularity. These polymers, consisting of linear molecules in which the chemical groups are regularly arranged along the macromolecule, are highly crystalline and consequently exhibit outstanding physical properties compared to their amorphous counterparts.

B. MOLECULAR STRUCTURE OF POLYPROPYLENE

Professor Natta defined this new type of polymer-forming process which produces these regular polymers as stereospecific polymerization, and the agents used to initiate such reactions were called stereospecific catalysts. The polymers—or the macromolecules comprising the polymers which possessed this structural regularity—are designated as "tactic," and structures with randomly positioned parts of the molecule as "atactic." Further, with reference to polypropylene specif-

1

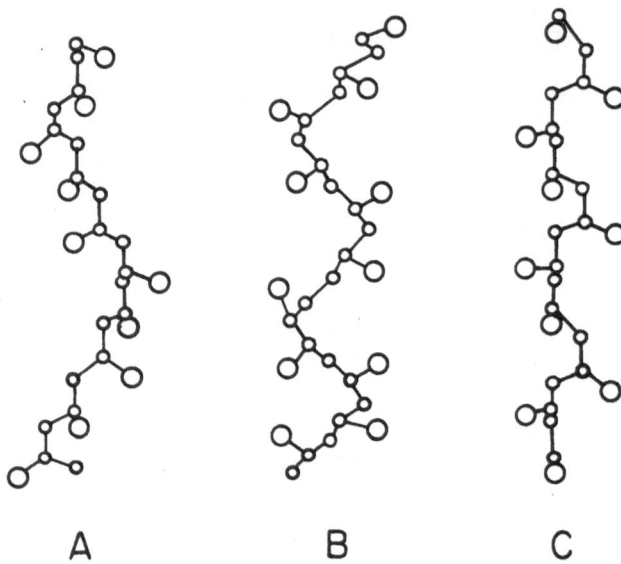

A   B   C

Figure 1. Molecular Configuration for Various Polypropylene
Structures [7]: (A) Atactic. (B) Syndiotactic. (C) Isotactic
(Left-Handed Helix).

ically, two highly crystalline, sterically regular structures are possible, defined as "isotactic" and "syndyotactic." In the isotactic arrangement, all the methyl groups are stationed on one side of the main chain of the polypropylene macromolecule; the methyl groups alternate regularly on both sides of the chain in the syndyotactic configuration. These two structural arrangements are shown below:

$$
\text{ISOTACTIC:} \quad
\begin{array}{cccccc}
H & CH_3 & H & CH_3 & H & CH_3 \\
| & | & | & | & | & | \\
-C- & C- & C- & C- & C- & C- \\
| & | & | & | & | & | \\
H & H & H & H & H & H
\end{array}
$$

$$
\text{SYNDYOTACTIC:} \quad
\begin{array}{cccccc}
H & CH_3 & H & H & H & CH_3 \\
| & | & | & | & | & | \\
-C- & C- & C- & C- & C- & C- \\
| & | & | & | & | & | \\
H & H & H & CH_3 & H & H
\end{array}
$$

The macromolecules of isotactic polypropylene conform to a helical shape of which four types exist, depending on the disposition of the methyl groups. The helix can be right-handed or left-handed, with the methyl groups "up" or "down." Chain structures of isotactic polypropylene with a left- and right-handed helix and atactic polypropylene are presented in Figure 1. Further, depending on the heat treatment of the polymer, two isotactic structures of polypropylene are possible: (1) a stable monoclinic form which is highly crystalline and (2) a less-crystalline supercooled "smetic" form. This paracrystalline, or smetic, form can be transformed to the monoclinic structure upon heating. X-ray diffraction patterns of these various polypropylene forms are given in Figure 2.

C. ADVANTAGES OF STEREOSPECIFIC POLYMERIZATION

An especially desirable aspect of stereospecific polymerization of propylene is realized when the many possible varieties of tacticity, as obtained by varying the conditions of polymerization, are considered. These varieties of tacticity

TABLE 1

CRYSTALLINE POLYMERIZATION MECHANISMS FOR
VARIOUS OLEFINIC MONOMERS

A. Monomers Which Produce Linear Crystalline Polymers by
Ordinary Processes of Polymerization

Monomers	Symmetrical monomeric units
Ethylene	H H | | - C — C - | | H H
Vinylidene chloride	H Cl | | - C — C - | | H Cl
Perfluorethylene	F F | | - C — C - | | F F

B. Monomers Which Produce Crystalline Polymers Only by Stereospecific Polymerization

Monomers	Asymmetrical monomeric units
Propylene	$\begin{array}{cc} H & CH_3 \\ | & | \\ -C & - C - \\ | & | \\ H & H \end{array}$
Styrene	$\begin{array}{cc} H & C_6H_5 \\ | & | \\ -C & - C - \\ | & | \\ H & H \cdot \end{array}$
Vinyl ethers	$\begin{array}{cc} H & OR \\ | & | \\ -C & - C - \\ | & | \\ H & H \end{array}$
Acrylates	$\begin{array}{cc} H & COOR \\ | & | \\ -C & - C - \\ | & | \\ H & H \end{array}$

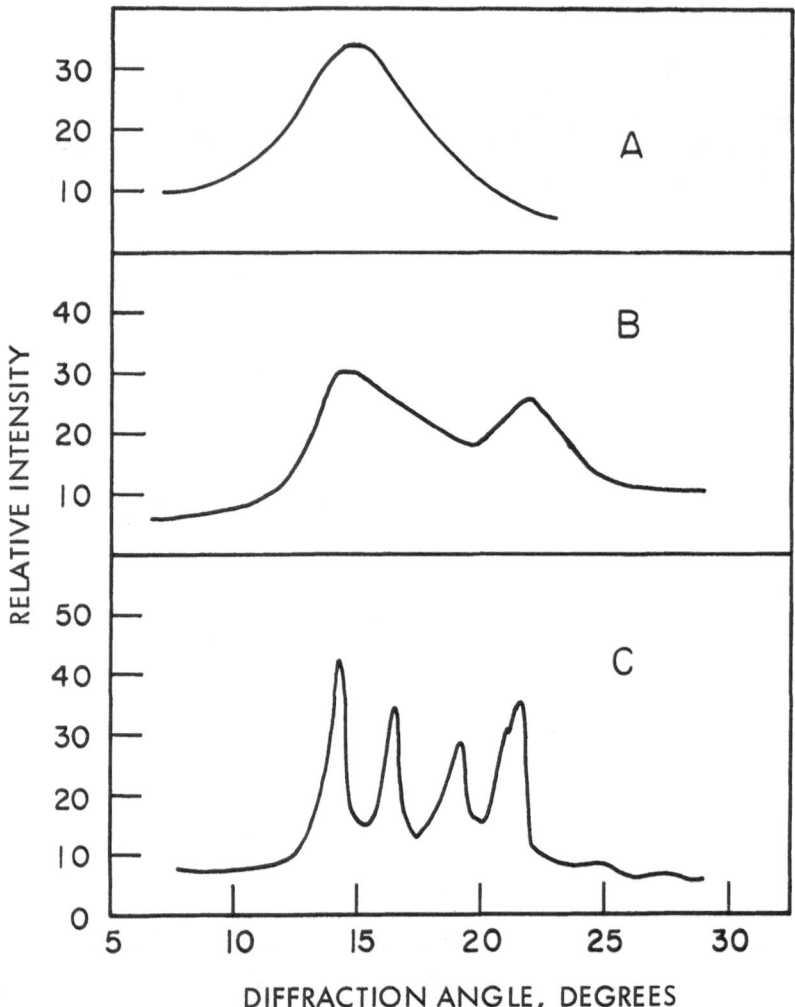

Figure 2. X-Ray Diffraction Patterns for Various Polypropylene
Structures [3]: (A) Atactic. (B) Isotactic ("Smetic"). (C) Isotactic (Monoclinic).

TABLE 2

COMPARATIVE COST OF VARIOUS MONOMERS AS
PRODUCED COMMERCIALLY IN THE UNITED STATES

Polymer	Cost of the monomer in the U. S., cents/pound
Polyethylene................	5–6
Isotactic polypropylene.........	3–5
Isotactic polybutylene..........	4–5
Isotactic polystyrene	13–14

stem from the numerous combinations of molecular weight, length variations of isotactic structural sections, and portions of atactic regions which comprise the polymers. This characteristic permits the preparation of polypropylenes with a wide range of mechanical properties from the same starting monomer, depending upon the steric structure imparted to the polymer. Thus a material can be tailored specifically, within relatively wide limits of physical properties, to best satisfy the needs of a particular application.

Many olefins have asymmetrical monomeric units and, as such, could not crystallize because of their structural non-uniformity. Table 1 denotes the polymerization mechanism required to crystallize various olefinic monomers.

Stereospecific polymerizations, by imparting a regular structure to the macromolecules, permit the use of simple and inexpensive monomers, factors of great importance in the over-all cost of the finished fibers. The cost factor of a new fiber in such a highly competitive industry as textiles is naturally of paramount importance. Table 2 shows comparative mono-mer cost data based on the production of monomers in the United States.

II. Fiber Manufacture Operations

Three basic methods of preparation of synthetic fibers are used commercially: (1) wet spinning, (2) dry spinning, and (3) melt spinning. In each process a viscous fluid is extruded through a multiholed die or spinneret, forming a fine-diameter fiber. Polypropylene fibers are prepared via the melt spinning technique, which essentially is comprised of two manufacturing stages: (1) extrusion of a fiber and (2) the subsequent thermal and mechanical stretching of the fiber. A diagram of a typical equipment line arrangement is shown in Figure 3. The various process equipment used in the preparation of polypropylene fibers is described below.

A. EXTRUSION

The melting of the resin, sometimes termed as "plasticating," is accomplished with a conventional thermoplastic extruder equipped with a polyethylene-type metering screw having a minimum 4:1 compression ratio and a metering zone no less than four flights in length. An extruder barrel with a length-to-diameter ratio of 24:1 is preferred since polypropylene requires higher extrusion temperatures than most other thermoplastic resins.

A wire cloth screen pack in the head of the extruder is positioned to prevent foreign particles from impregnating the extruded fibers. Since highly oriented fibers, such as polypropylene, are sensitive to contamination breakage, this screening is expedient. A pressure control valve is generally in-

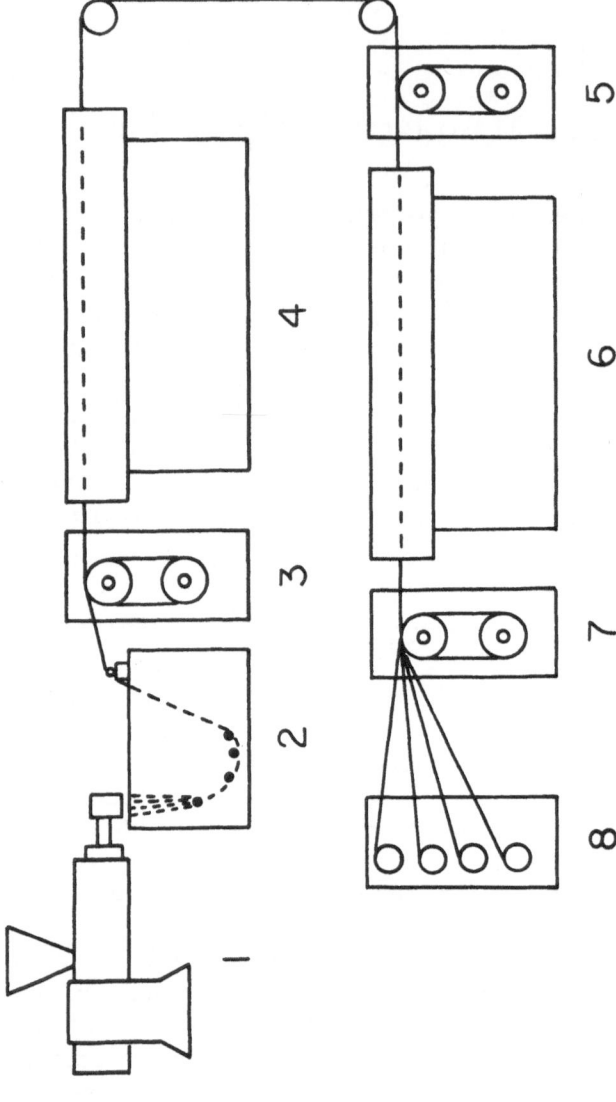

Figure 3. Typical Line Diagram for the Extrusion of Polypropylene Monofilaments [28]: (1) Extruder. (2) Quench Tank. (3) Pull–Out Rolls. (4) Draw Oven. (5) Draw Rolls. (6) Relax Oven. (7) Relax Rolls. (8) Wind–Up.

stalled after the screen pack (immediately upstream of the die) to compensate for variations in polymer throughputs at a given screw speed, which may result from varying polymer melt viscosity or barrel temperature profiles. This pressure control is best obtained at the 90° bend of the polymer flow stream inherent in monofilament die heads; the die head is essentially the transition section of the extrusion line which conveys the extruded melt to the die.

The die containing the capillaries through which the molten polymer is forced into fibers is mounted to the downstream side of the die head. The melt temperature and pressure are monitored at the entrance to the die by thermocouples and pressure gauges, or, preferably, pressure transducers. The extrusion die is the most important single part of the monofilament operation and requires careful design, machining, and maintenance. Filament irregularities which can be attributed to deficiencies at the die include (1) end-to-end fiber diameter variation, (2) surface flaws, (3) flow pulsation and attendant diameter variations in the fiber produced, and (4) excessive fiber waviness in the quench bath.

B. DRAW-DOWN, QUENCHING, AND DRYING

Between the die exit and the pull-out rolls (first Godet station), the extruded fibers are predrawn, water-quenched, separated into an orderly arrangement, and dried of adhering surface moisture.

The predraw, or draw-down, is a decrease in fiber diameter which occurs as the molten polymer emerges from the die; the diameter of this unoriented fiber is approximately 50% of the die hole diameter. The operational variables controlling the uniformity of draw-down are polymer throughput and filament velocity just prior to the quench bath. Upon solidification of the fiber in the quench bath, draw-down is complete.

Fast quenching retards crystal growth and results in an amorphous fiber composed of a large number of small crystallites; this type of fiber excels in toughness and flexibility.

Quenching at higher bath temperatures promotes crystal growth, resulting in a more crystalline fiber which exhibits superior strength and rigidity.

The quenched fibers are drawn around an adjustable guide assembly located in the quench bath and over another set of guide bars positioned at the exit of the bath, which serves to maintain an orderly arrangement of the fibers. An air jet is directed at the "tow" to surface-dry the fibers, upon their exit from the bath. Removal of this moisture is of prime importance as uneven heating will result in the fiber orientation section.

C. ORIENTATION, RELAXATION, AND WIND-UP

The fibers then pass into the orientation, or draw, stages, which impart thermal and mechanical treatments to the fibers. Initially the temperature of the fibers is raised by an orientation oven in preparation of the all-important stretching of the fibers. This step is the most single important operation in the manufacture of fibers relative to the properties exhibited by the fibers. The stretching of the fibers is effected by high-speed draw rolls driven by powerful motors; the ratio of the speed of the pull rolls and the draw rolls is the draw ratio. The higher the draw ratio, the more orientation imparted to the fibers.

In order to minimize shrinkage, by relieving residual stresses, the fibers are then permitted to relax and become heat-set by passing through another oven. The heat-set temperature must be higher than the end use of the fibers in order to be effective in preventing shrinkage. The last set of rolls which pulls the fibers through the relax oven is driven at a speed slightly lower than that of the draw rolls. The individual fibers in the tow are then wound upon a spindle in the wind-up station. This wind-up unit is generally synchronized with the speed of the other rolls to maintain uniform winding.

III. Influence of Fiber Processing Conditions on the Structural and Physical Properties of Polypropylene Filaments

The properties of polypropylene fiber are dependent upon the structural characteristics of the macromolecules comprising the polymer; in turn, these macromolecules are significantly affected by (1) polymer properties prior to extrusion, and (2) processing conditions of fiber formation.

The pertinent structural characteristics of a polymer include molecular weight, average molecular weight (distribution), crystalline structure, crystallinity, orientation, and terminal end groups. The significant manufacturing conditions are extrusion temperature, draw ratios, annealing time and temperature, and quenching time and temperature. Whereas the crystallinity and orientation are affected by quenching, drawing, and annealing conditions, the extrusion conditions affect the crystallinity, orientation, and molecular-weight distribution.

A. GENERAL CONSIDERATIONS OF CRYSTALLINITY, ORIENTATION, AND MOLECULAR WEIGHT OF FIBERS

One of the most important polymer characteristics with respect to determining the chemical and physical properties of fibers is crystallinity. All fibers consist of macromolecules

TABLE 3

CONTRIBUTIONS OF THE CRYSTALLINE AND AMORPHOUS
AREAS TO THE CHARACTERISTICS OF FIBERS

Crystalline areas	Amorphous areas
Stiffness.	Plasticity
Moisture stability	Absorbency
Temperature stability.	Formability
Dimensional stability	Dyeability
Network strength.	Toughness

which have extremely long lengths relative to their diameters. Only those fiber crystals containing macromolecules which are sterically and chemically regular ("ordered-arrangement") exhibit any appreciable degree of crystallinity; the larger the crystalline areas in the polymer, the higher the degree of crystallinity.   In turn, the crystals can also be arranged in a well-ordered geometry with respect to the fiber axis, an arrangement which results in the longitudinal stretching of the fiber during manufacture.   This ordered alignment of the polymer crystals along the fiber axis is a measure of orientation.   Table 3 qualitatively presents the contributions of the amorphous and crystalline areas to the characteristics of a fiber, and indicates that the strength, stability, and stiffness of a fiber are attributed to the crystalline area, while the amorphous region is responsible for the fiber workability characteristics.

Another parameter which exerts a strong influence on the properties of fibers is the molecular weight of the polymer. Characteristics of high-molecular-weight materials are long molecular chains which are comprised of many crystals. This is in contrast with the large number of complete molecules within the crystals of low-molecular-weight substances. As a result of this crystal-binding action, high-molecular-weight fibers exhibit good mechanical properties.

B. INFLUENCE OF FIBER PROCESS CONDITIONS ON THE STRUCTURAL CHARACTERISTICS OF POLYPROPYLENE FIBERS

1. CRYSTALLINITY

a. Effect of Polymer Melt Temperature on Crystallization Rate

Crystallization of a polymer is the result of two processes: (1) nucleation and (2) nuclei growth.   The rates of both nucleation and growth processes are dependent upon the crystallization temperature as well as on the temperature

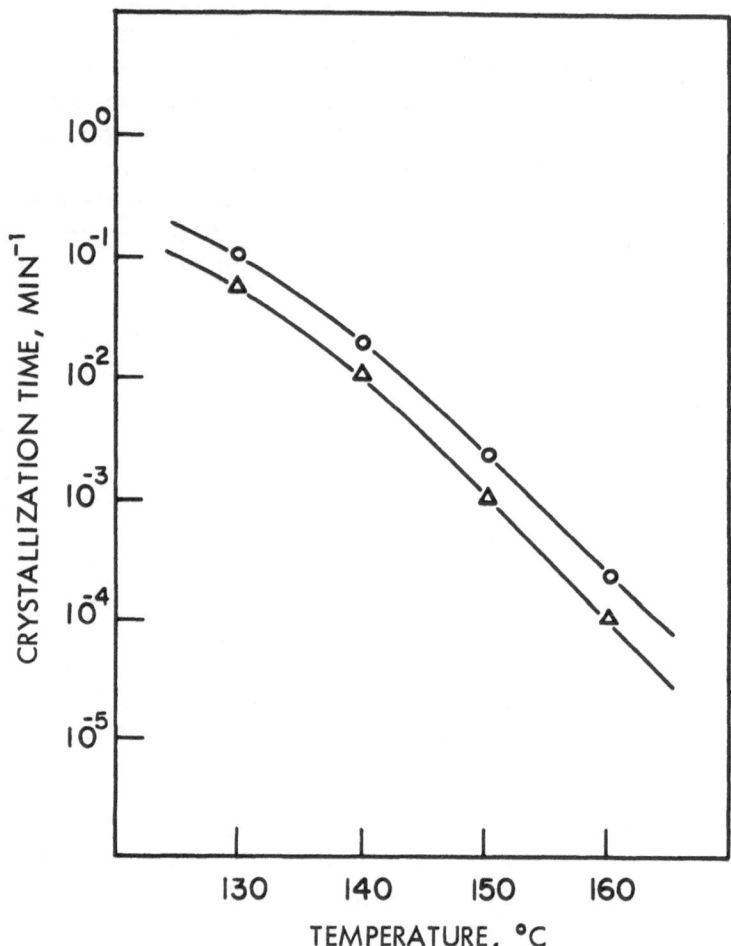

Figure 4. The Effect of Polymer Temperature on the Crystallization Rate of Polypropylene [3]: (O) Polymer Heated at 200°C
for 30 min. (Δ) Polymer Heated at 240° for 30 min.

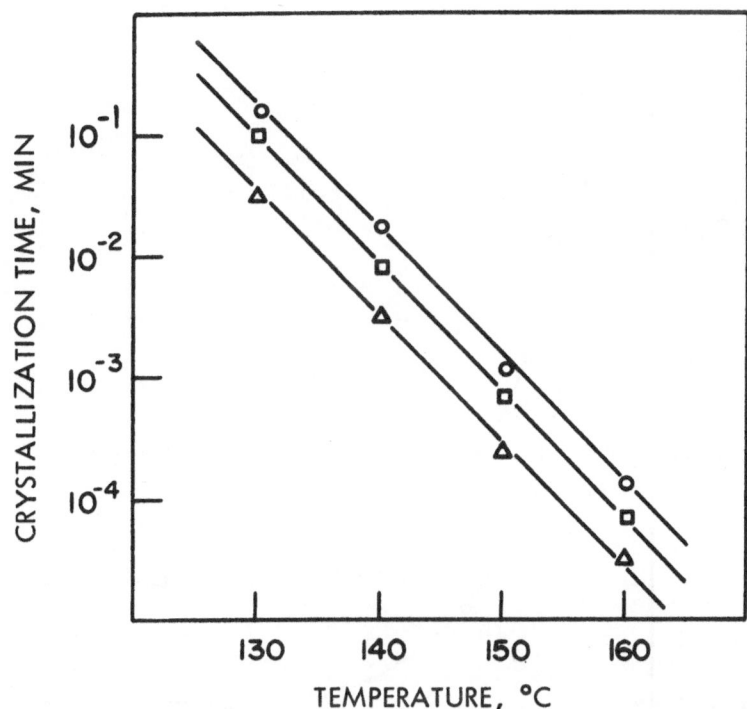

Figure 5. Relationship Between Rate of Crystallization and Temperature for Various Molecular-Weight Polypropylenes [7]: [Molecular Weight ($\bar{M}_w$)] (O) 173,000. (□) 357,000. (Δ) 490,000.

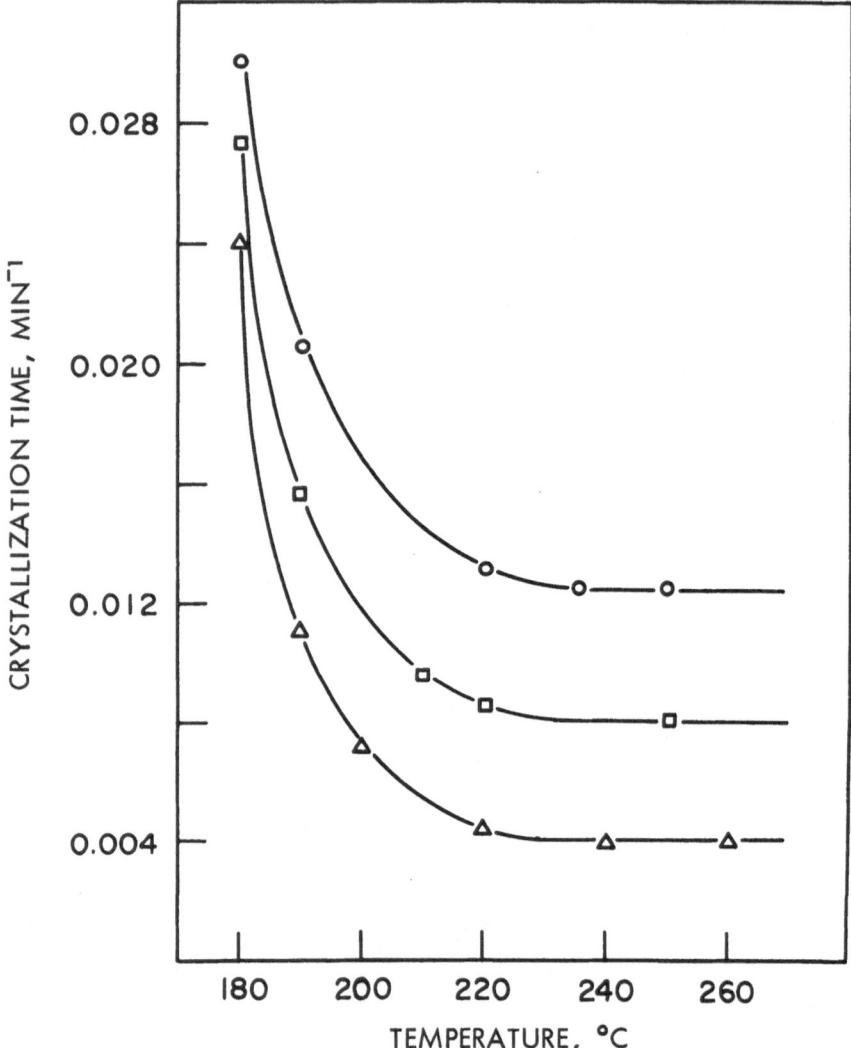

Figure 6. The Crystallization Rate of Polypropylene as a Function of Polymer Melt Temperature and Molecular Weight [7]:
[Molecular Weight ($\overline{M}_w$)] (○) 173,000.
(□) 357,000. (△) 490, 000.

and duration of the molten polymer state. Figure 4 illustrates the dependence of the rate of crystallization on the crystallization temperature for molten polypropylene at 200 and 240°C. The investigators used the reciprocal value of the time required to attain one-half of the equilibrium crystallinity as a measure of crystallization speed. The plot shows that the crystallization rate decreases sharply with increased crystallization temperature. As expected, the molecular weight of the polymer influences the rate of crystallization; this influence is manifested by a slight increase in the rate with decreasing molecular weight polymers as shown in Figure 5.

The dependence of the crystallization speed of three polypropylenes of different molecular weights as a function of temperature is shown in Figure 6. The polypropylene melts were maintained for 1 hr at the crystallization temperature of 140°C. The plot shows that the rate of crystallization decreases as the melt temperature is increased to about 200°C; the crystallization rate is unaffected by further increase in temperature. This plot also illustrates the decrease in crystallization rate with increased polymer molecular weight; when the temperature is constant, the rate of crystallization depends on molecular weight as well as on any foreign substances, such as stabilizers, added to the polymer.

b. Effect of Quench Temperature on the Crystalline
Structure of Polypropylene Fibers

In addition to the effect of spinning temperature on crystallization as described above, crystallinity is also significantly influenced by the temperature and rate of fiber cooling after extrusion, i.e., the quenching conditions. When polypropylene fibers are quenched immediately after extrusion at a low bath temperature, the result is formation of the amorphous "smetic" type of polypropylene, a supercooled paracrystalline structure with a low crystallinity level; in contrast, quenching at a high bath temperature or further downstream of the die (slow cooling) promotes crystal growth, which results in fibers with much higher crystallinity than that of the smetic structure.

Quantitative data on the relationships between structure and crystallinity are shown in Table 4 for fibers quenched

TABLE 4

PHYSICAL PROPERTIES OF POLYPROPYLENE FILAMENTS
PREPARED BY EXTRUSION METHODS A AND C AND
QUENCHED UNDER DIFFERENT CONDITIONS[a]

Quenching conditions	Crystalline structure	Crystallinity of undrawn filaments, %
Extrusion method A, melt temperature, 280°C		
Water bath, 50°C	Unoriented paracrystalline	44
Air, 20°C	Slightly oriented crystalline	54
Extrusion method C, melt temperature, 235°C		
Water bath, 10°C	Unoriented paracrystalline	45
Air, 20°C	Slightly oriented crystalline	62

[a] The above fibers were prepared from Polymer No. 6723 (molecular weight of 350,000 – 470,000) from the Hercules Powder Company.

in air and in water. The fibers were spun from "Pro-fax" polypropylene polymers under two sets of extrusion conditions, "A" and "C," described as follows:

Method A: The polymer was extruded at 280°C before which no attempt was made to remove the air from the polymer or extrusion pot.

Method C: The fibers were extruded at 235°C with a minimum exposure to air; the polymer and extrusion pot were air-evacuated (from 5 to 10 mm Hg) prior to the extrusion of the fibers.

The table shows that the water bath produces the amorphous structure while an air quench yields the crystalline form. The highest crystallinity of the air-quenched fibers prepared by method C (62%) is the result of the minimum difference between the extruded fiber temperature and the quench temperature; the crystallinity is lowered to 54% for the air-quenched Method-A fibers extruded at 280°C. The lowest crystallinity of 44–45% is obtained with Method-C fibers quenched in water (fastest quench).

It should be repeated here that the X-ray diffraction pattern of the smetic form, although typical of amorphous polymers having low crystallinities, is not identical to the atactic polypropylene structure as shown in Figure 2. Upon heat treatment, this smetic form is converted into the stable monoclinic structure, a factor of great importance in the preparation of high-tenacity polypropylene fibers, as will be discussed in a subsequent section.

c. Effect of Fiber Draw Rate and Draw Temperature on Crystallinity

In addition to the above influence of polymer melt and fiber quench temperatures on crystallinity, fiber draw ratio and draw temperature are other variables which significantly affect the crystallization of polypropylene fiber. Although the primary purpose of fiber draw is to orient the crystallites of the polymer, the stretching operation also decreases the crystallinity of polypropylene fibers as shown in Figure 7.

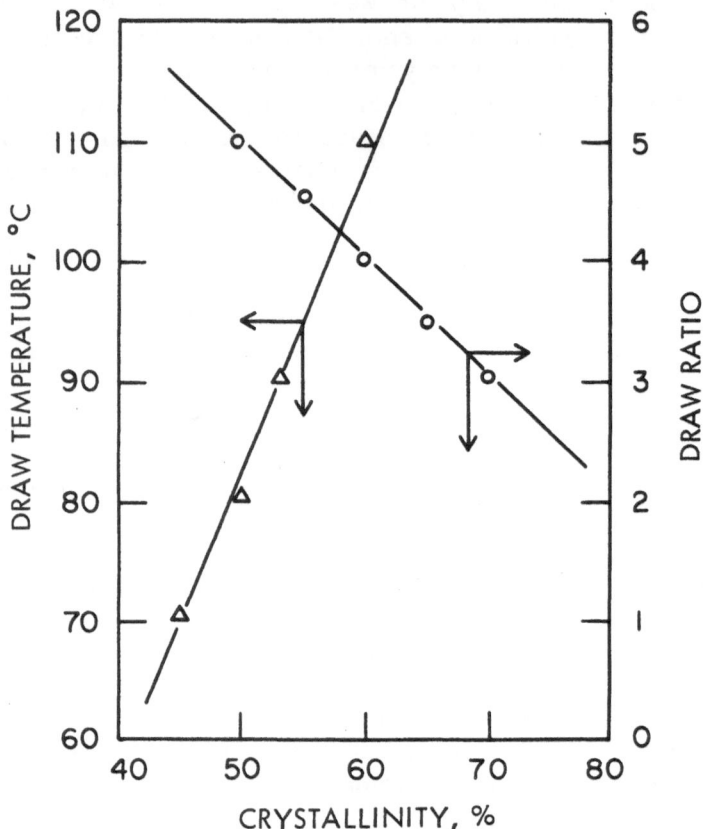

Figure 7. Effect of Draw Ratio and Draw Temperature upon
Polypropylene Fiber Crystallinity [3]: (○) Draw Temperature
of 100°C. (△) Draw Ratio of 4:1.

As expected, the decrease in crystallinity upon fiber draw is less at higher draw temperatures because the oriented chains tend to crystallize; this relation is also shown in Figure 7. At high draw ratios, of about 25:1, a balance between the changes in crystallinity as affected by fiber draw (decrease) and temperature (increase) is reached as shown in Figure 8.

2. MOLECULAR WEIGHT

Some of the relationships between extrusion conditions and fiber molecular weight have been investigated by the Southern Research Institute. The variation of fiber molecular weight and molecular-weight distribution of polypropylene filaments as found in this study is presented in Table 5. The fibers were spun under two sets of extrusion conditions, A and C, as described in the previous section.

The molecular weight of the polymers studied ranged from 245,000 to 470,000 as calculated from intrinsic viscosities of the polymers in decalin; the ratio of the weight-average molecular weight to the number-average molecular weight, $\overline{M}_w/\overline{M}_n$, as reported by the polymer manufacturer was 10.0. From the tabulated data of Table 5, extrusion in the presence of air (method A) results in a 70–77% reduction in polymer weight-average molecular weight, while in sharp contrast, the decrease is only 3–16% for air-free extrusions. Because of the low thermal degradation of the polymers extruded under Method C, the molecular-weight distribution of the filaments decreased slightly (8.5–9.3) from the original value of 10.0. Filaments prepared by extrusion method A suffered extensive thermal degradation as evidenced by the narrow-molecular-weight distribution range of 2.1–3.5.

3. ORIENTATION

In addition to exerting a strong influence on crystallinity, quench conditions, along with draw ratio, determine the degree of fiber orientation. Under conditions of slow cooling, orientation occurs that is parallel to both the fiber and to the direction of extrusion, resulting in a cross-linked fiber.

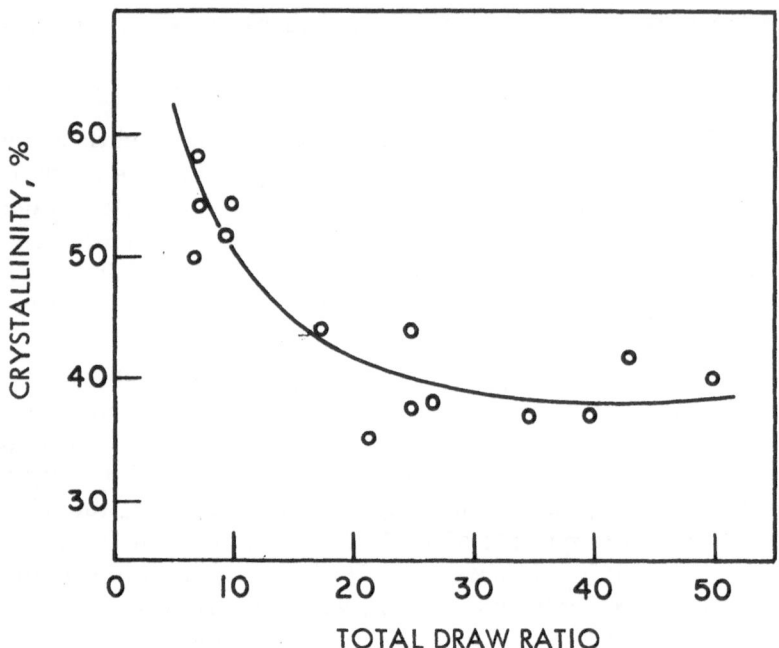

Figure 8. Relationship Between Draw Ratio and Crystallinity
of Annealed Polypropylene Filaments [34].

TABLE 5

EFFECTS OF EXTRUSION CONDITIONS ON THE
MOLECULAR WEIGHT AND RANGE OF THE MOLECULAR-
WEIGHT DISTRIBUTION OF THE POLYPROPYLENES
TESTED

Polymer	Extrusion method	Molecular weight, $\bar{M}_w$ [a]		Range of fiber molecular-weight distribution, $\bar{M}_w/\bar{M}_n$ [b]
		Polymer	Filament	
6323	A	245,000	67,000	2.5
	C		225,000	9.1
6423	A	285,000	78,000	2.1
6523	A	305,000	88,000	2.1
6623	A	315,000	92,000	3.5
	C		305,000	9.3
6723	A	470,000	110,000	2.1
	C		405,000	8.5

[a] The molecular weights were determined from intrinsic viscosities of the materials in decalin.
[b] $M_w/M_n = 10 \left([\eta]/[\eta]_0\right)^{1.25}$ where $[\eta]_0$ is the intrinsic viscosity of the polymer before extrusion and $[\eta]$ is the intrinsic viscosity of the polymer after extrusion.

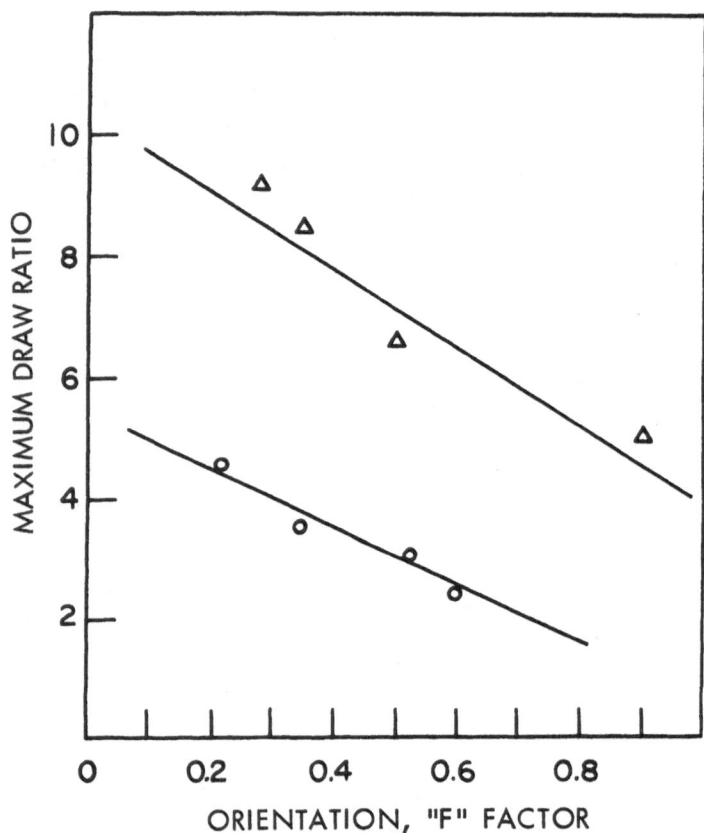

Figure 9. Maximum Draw Ratio as a Function of Orientation
for Amorphous and Crystalline Polypropylene Fibers [7]:
(Δ)Amorphous . (O) Monoclinic (50–60% Crystallinity).

Slight orientation of the chains parallel to the drawing direction occurs only upon fast quench conditions. By proper selection and control of extrusion temperature, quench time and temperature, and draw conditions, a wide range of orientation and crystallinity in polypropylene fibers may be obtained.

Fibers extruded under process conditions which yield a crystalline structure (slow quench) are much more resistant to orientation than fibers with the smetic amorphous structure at equal temperatures and drawing rates; the ease of stretching the amorphous structure is attributed to the greater mobility of the macromolecules in this form of polypropylene. This relationship of orientation and crystallinity levels is shown by Figure 9, which presents the maximum draw ratio as a function of orientation for the amorphous and crystalline types of polypropylene fiber. The investigators calculated the degree of fiber orientation by the equation $f = \frac{1}{2} (3\cos^2\theta - 1)$ where θ is the angle between the axis of molecules and the direction of stretching; the degree of orientation is inversely proportional to the angle, i. e., the higher f values represent lower orientation. These plots show that, over the range of draw ratios investigated, the degree of orientation for a given maximum draw ratio is about 3—4 times greater in the amorphous structure relative to the crystalline sample.

The data of another investigation which explored the effects of high draw ratios (up to 50:1) on the degree of orientation are presented in Figure 10. In this plot the degree of orientation is expressed in terms of the half-mean width of the diffraction ring of the X-ray pattern on the 110 plane; as above, the smaller angles represent higher degrees of orientation. These results agree with those of Figure 9 discussed above in that the smetic amorphous structure orients more rapidly than the crystalline-type fiber; in addition, Figure 10 shows that both structural types become completely oriented at a total draw ratio of about 25:1.

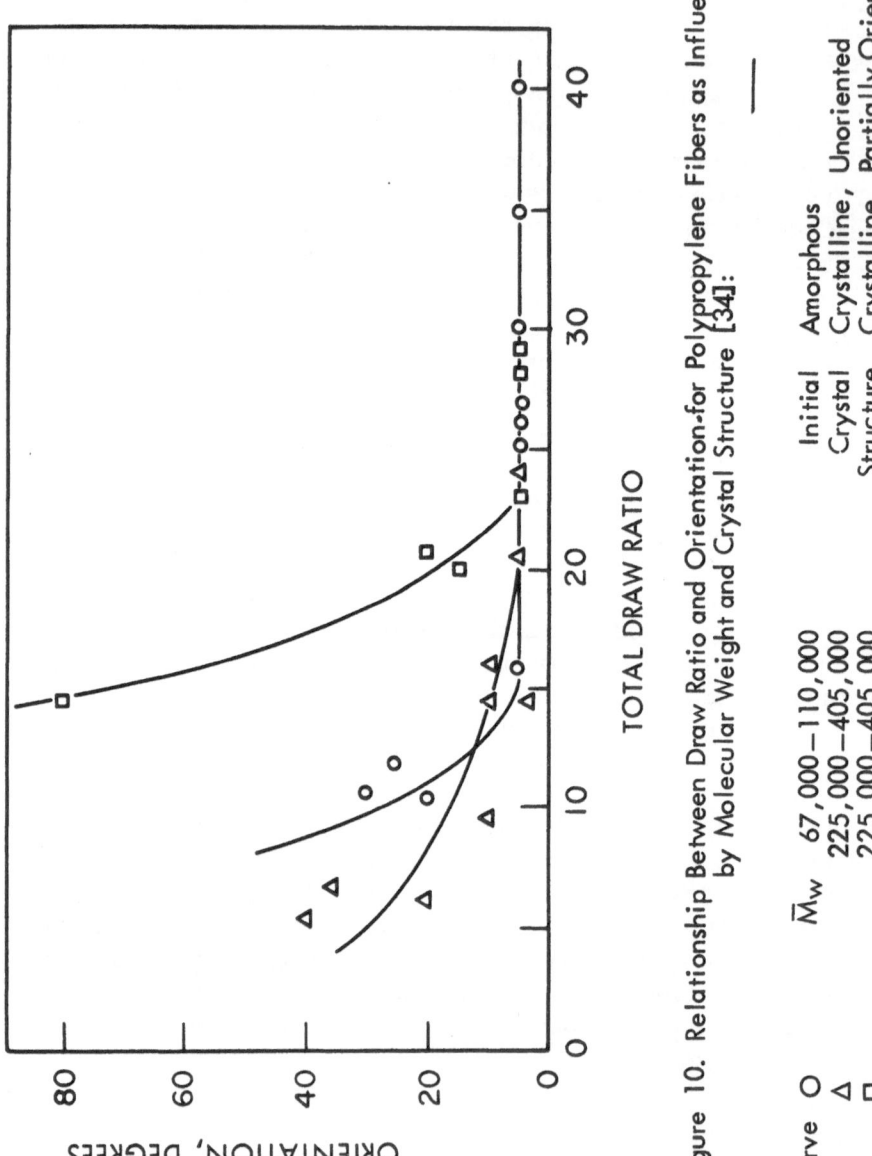

Figure 10. Relationship Between Draw Ratio and Orientation for Polypropylene Fibers as Influenced by Molecular Weight and Crystal Structure [34]:

Curve	$\bar{M}_w$	Initial Crystal Structure
O	67,000–110,000	Amorphous
△	225,000–405,000	Crystalline, Unoriented
□	225,000–405,000	Crystalline, Partially Oriented

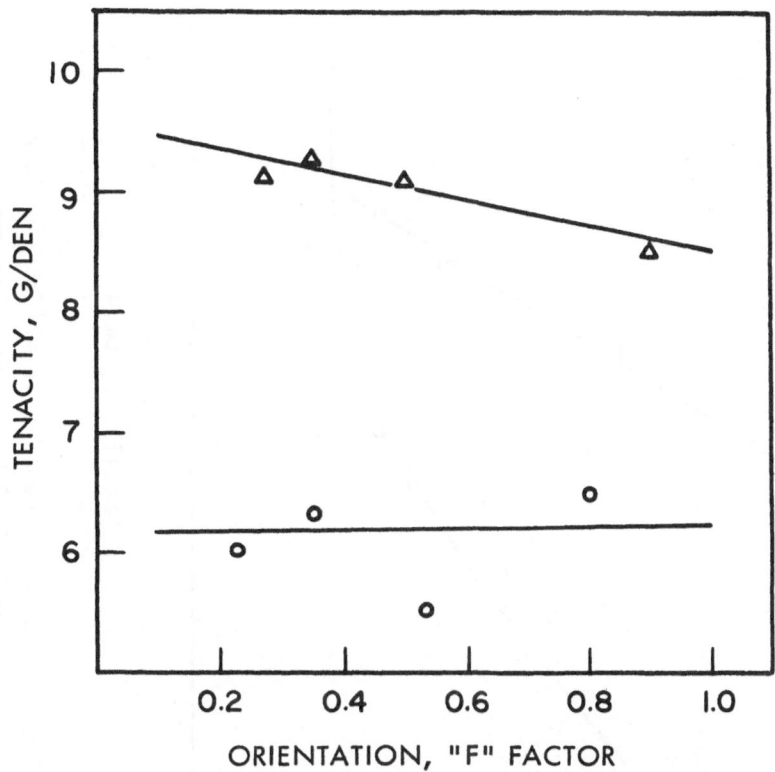

Figure 11. Variation of Tenacity with Orientation for Amorphous and Crystalline Polypropylene Fibers [7]: (Δ) Amorphous. (O) Monoclinic (50-60% Crystallinity).

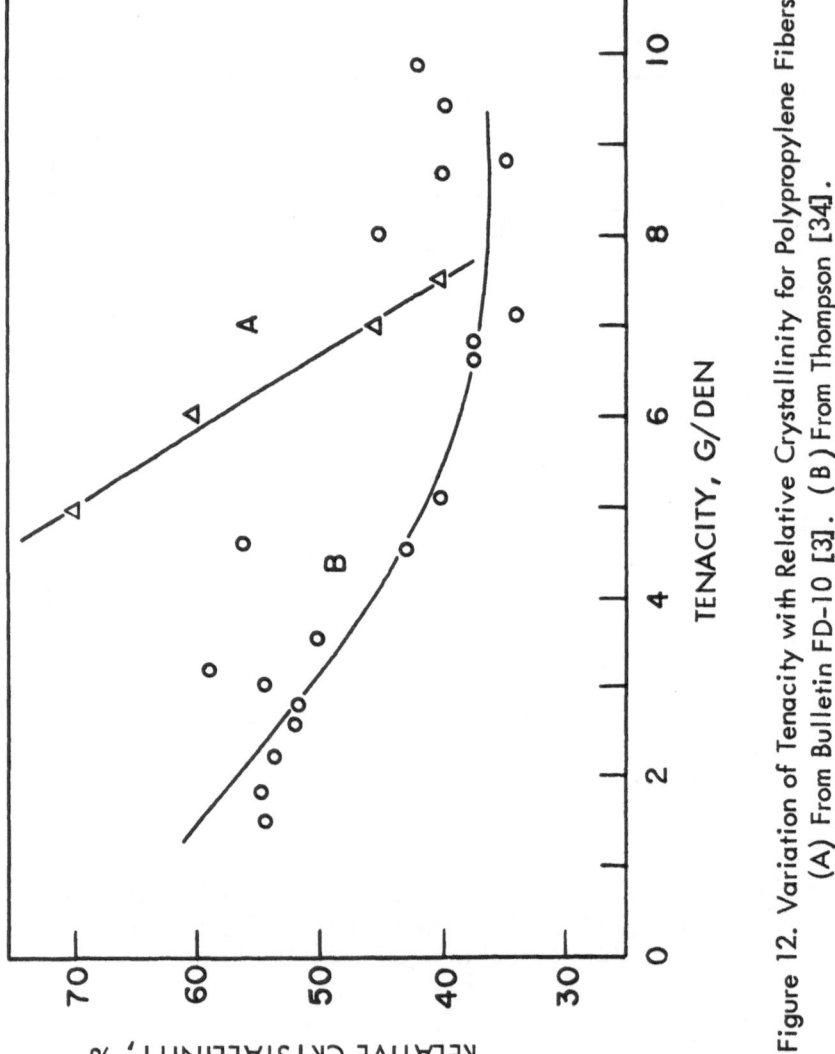

Figure 12. Variation of Tenacity with Relative Crystallinity for Polypropylene Fibers:
(A) From Bulletin FD-10 [3]. (B) From Thompson [34].

C. EFFECTS OF POLYMER STRUCTURAL PARAMETERS AND FIBER PROCESS CONDITIONS ON THE PHYSICAL PROPERTIES OF POLYPROPYLENE FIBERS

1. RELATIONSHIPS AMONG CRYSTALLINITY, ORIENTATION, AND PROPERTIES OF POLYPROPYLENE FIBERS

As mentioned previously, the crystallinity level of an extruded polypropylene fiber is highly dependent on the quench history of the molten polymer; in turn, the subsequent orientation of these fibers is inversely related to the imparted degree of crystallinity. Consequently, the supercooled, low-crystalline structure of polypropylene is most easily oriented.

Since fiber tenacity increases with fiber orientation, the amorphous forms of polypropylene show higher tensile strengths than the more crystalline modifications at practical levels of drawing ratios; this variation of tenacity with orientation for both the amorphous and monoclinic polypropylene fiber structures is shown in Figure 11. As seen from these plots, fiber tenacities are about 9 g/den and 6 g/den for the amorphous and crystalline samples, respectively, over the range of orientation investigated. [The cross-sectional degree of orientation in Figure 11 was calculated with the equation: $f = \frac{1}{2}$ $(3 \cos^2 \theta - 1)$ as described earlier.] As indicated by Figure 10 discussed earlier, at sufficiently high draw ratios (above 25) the orientation of polypropylene fibers becomes independent of polymer crystalline structure. Consequently, at these extraordinary draw ratios, the tenacities of drawn polypropylene fibers are independent of the crystallinity of the undrawn filaments.

The interrelationship among crystallinity, orientation, and fiber tensile strength is further illustrated by Figure 12, which presents data from two separate investigations. Plot B of the figure shows that as the crystallinity decreases from about 60% to 40%, the tenacity increases from 1 g/den to about 6 g/den; below 40%, crystallinity exerts no further influence on the tenacity. Cappuccio et al. did not investigate the effect

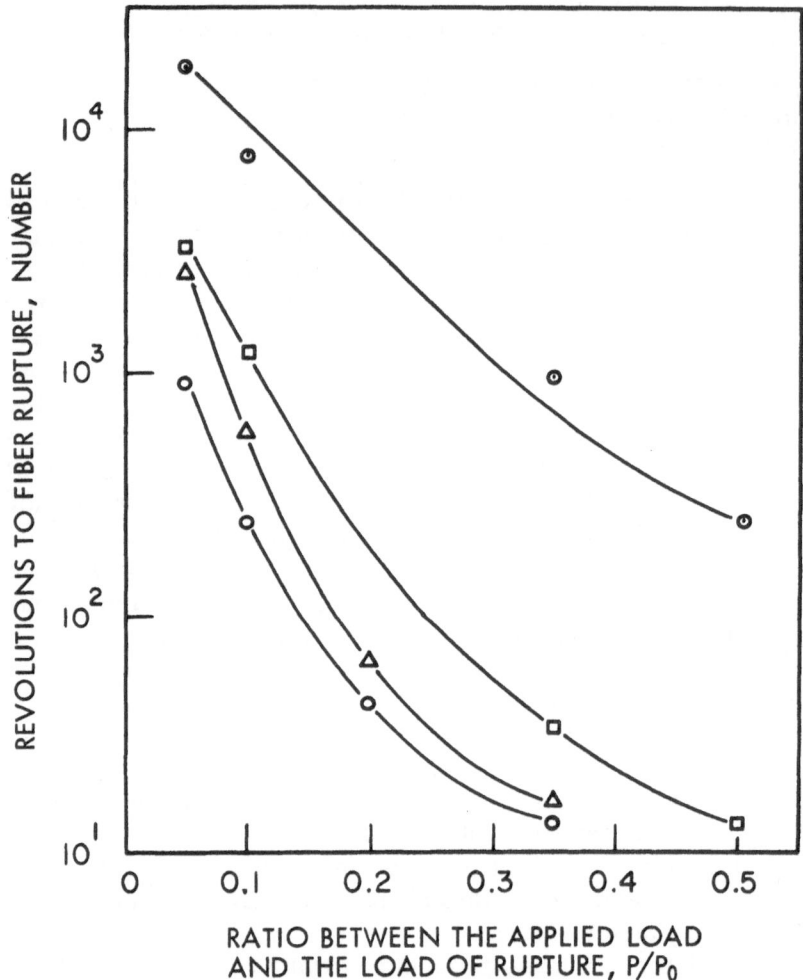

Figure 13. Effect of Crystallinity of Polypropylene Fibers upon
Abrasion Resistance [3]: (⊙)88%Crystallinity. (□)69%Crystal-
linity. (△) 50%Crystallinity. (O) 30% Crystallinity.

of crystallinity below 40%; consequently, they did not observe the independent phase of tenacity and crystallinity (plot A).

The crystallinity of a polymer extends its influence on the properties of a fiber to include permeability of liquids also. The solvent penetration suffered by partially crystalline materials occurs in the amorphous areas of the polymer molecular structure. For applications requiring exposure to severe chemicals, fibers of the highest crystallinity are specified to minimize swelling or even deterioration. The significance of permeability upon fiber properties will be discussed in the section which treats the chemical resistance of fibers.

The resistance to abrasion is a prime consideration in determining the usefulness of fibers in the textile industry. The influence of crystallinity upon abrasion resistance is illustrated by the plot in Figure 13. The data were obtained by abrading the fibers with a rotating carborundum cylinder until rupture; the number of revolutions required to fracture the fiber is plotted as a function of the ratio between the applied load and the load to rupture. The plot shows that a significant increase of abrasion resistance results in increased crystallinity.

2. EFFECT OF POLYMER MOLECULAR WEIGHT ON POLYPROPYLENE FIBER PROPERTIES

The effects of molecular weight on the properties of polypropylene fibers have been investigated by the Eastman Kodak Company and the Southern Research Institute. The information presented in this section is derived primarily from these works.

In the study conducted by the Eastman Kodak Company, polymer "flow rate" was utilized as a measure of molecular weight, the relationship being inversely proportional, i.e., a high-molecular-weight material exhibits a low flow rate. The values of flow rate used in this test were based on the weight of polymer flow, in grams, for a 10-min period at a temperature of 230°C and a load of 2.16 kg. The various molecular-

TABLE 6

POLYPROPYLENE FIBER PROCESSING CONDITIONS FOR THE EASTMAN KODAK STUDY

Formula number	4222	4232	4242	4252	4262
Melt temperature range, °F	428–536	410–572	392–536	374–590	356–500
Orientation temperature, °F	325	300	300	275	275
Orientation draw ratio	11:1	11:1	11:1	11:1	11:1
Denier, g/9000 m	360	460	430	510	440

weight "Tenite" polypropylenes tested were types 4222, 4232, 4242, 4252, and 4262, which represent flow rates of 2.5, 4.5, 9.0, 18.0, and 35.0 g/10 min, respectively. The relationship between polypropylene molecular weight and flow rate as determined under the above extrusion conditions is illustrated in Figure 14.

The fiber process conditions and properties are tabulated in Table 6 for the various flow-rate polypropylenes tested. Screw speed, quench bath temperature, and draw ratio were held constant for all tests, while melt and draw temperatures were altered as necessary to maintain the specified draw ratio for each type of flow-rate material.

The variation in the following properties were investigated: denier, Uster evenness, tenacity, elongation, modulus, elastic recovery, hot-water and hot-oven shrinkage, and filament shape. It should be noted that all elongation and tenacity data were obtained on an Instron tensile machine. The measurement of fiber properties is markedly influenced by the testing equipment used, the values obtained on an Instron tensile machine being consistently lower compared to those recorded on other tensile testers.

Denier. At the same spinning conditions, fiber denier increases with increasing flow-rate polymers; this effect is the result of the reduction of polymer viscosity in the higher-flow-rate polypropylenes.

Uster Evenness. The insignificant variation in filament size as measured by an Uster Evenness Tester indicates that this fiber property is not affected by polymer molecular weight but is highly dependent upon processing conditions.

Tenacity, Elongation, and Modulus. These properties represent fiber strength, increase in length, and rigidity, respectively. The relation of these properties with flow rate is plotted in Figure 15, which shows that these properties decrease as the flow rate increases. The low-flow-rate materials, corresponding to the highest-molecular-weight polymers, form the strongest and most rigid fibers.

Elastic Recovery. The elastic recovery of a fiber is the degree of its original length a fiber recovers after removal

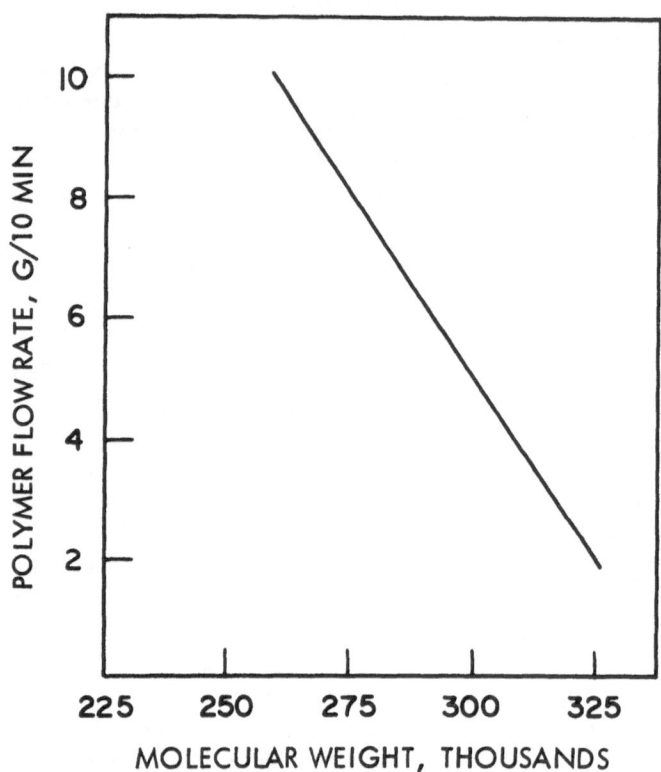

Figure 14. Relationship Between Molecular Weight and Polymer Flow Rate for Polypropylene [11,34].

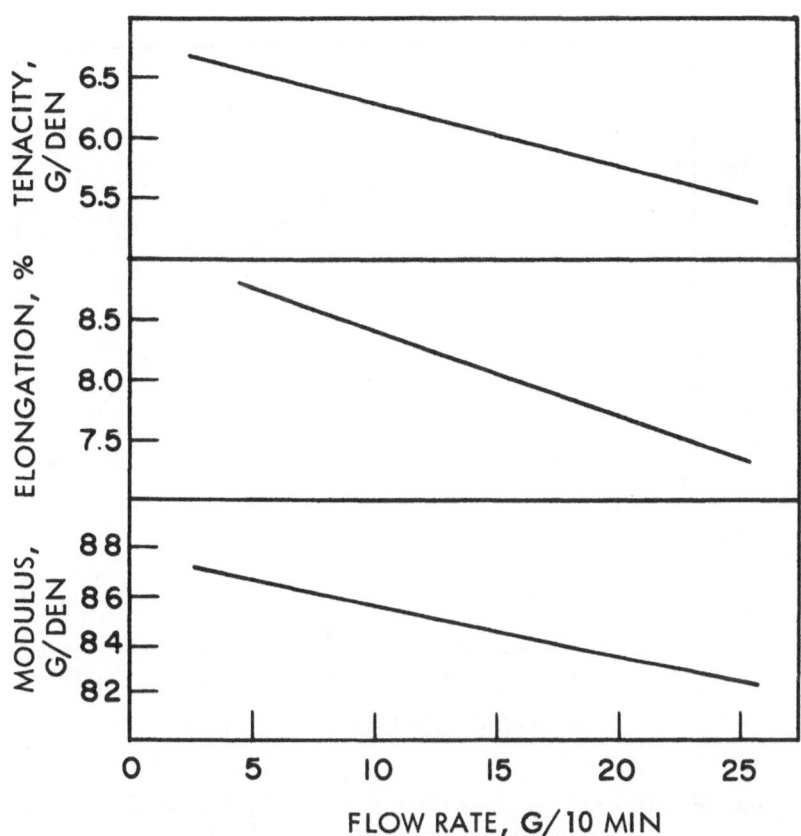

Figure 15. Effect of Polymer Flow Rate on Tenacity, Elongation, and Modulus of Polypropylene Monofilaments [11].

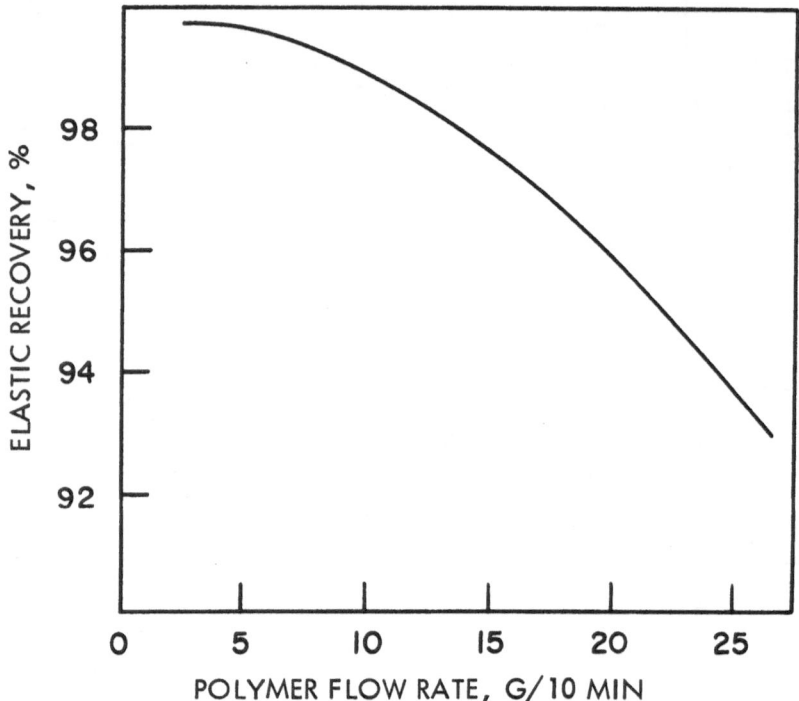

Figure 16. Effect of Polymer Flow Rate on the Elastic Recovery
of Polypropylene Monofilaments [11].

of a load which caused a specified elongation. In addition to the amount of extension, the relaxation time—the time allowed for the fiber to recover from the applied load—must also be specified. In this test the fibers were extended 4 and 8% with a relaxation time of 1 min. The fibers subjected to 4% extension showed complete recovery of their original length for the high- and low-flow-rate materials tested. However, at the higher percent elongation, the recovery properties of the fiber are dependent upon the flow rate of the material as shown in Figure 16. Only the fibers prepared from the lowest flow-rate material (2.5 g/10 min) exhibit 100% recovery of their original length after an 8% elongation. The nonelasticity of the lower-molecular-weight fibers after an 8% elongation is illustrated in the plot by the failure of these fibers to recover 100% of their original length.

Hot-Oven and Hot-Water Shrinkage. The variation of hot-oven and hot-water shrinkage with flow rate is presented in Figure 17 which illustrates that fibers prepared from higher-molecular-weight polymers are more resistant to thermal shrinkage than low-molecular-weight fibers. This property is attributed to the proximity of the draw temperature used for these fibers to the hot-oven and hot-water test temperatures; these draw temperatures impart a heat-set to the fibers which is destroyed only by subsequent exposure to temperatures considerably higher than this draw temperature. To maintain the constant draw ratio for all the fibers tested, the draw temperatures used for the higher-flow materials was necessarily lower, which accounts for their lower thermal stability.

Filament Shape. Polymer flow rate and melt temperature effects upon the shape of polypropylene fibers were found to be interrelated. Round, uniform fibers are prepared from high-flow materials over the range of melt temperatures employed, while fibers extruded from low-flow polymers are oval-shaped. Fibers extruded from the intermediate-molecular-weight polymers at high melt temperatures are round, but are changed to oval-shaped fibers with low-temperature melts.

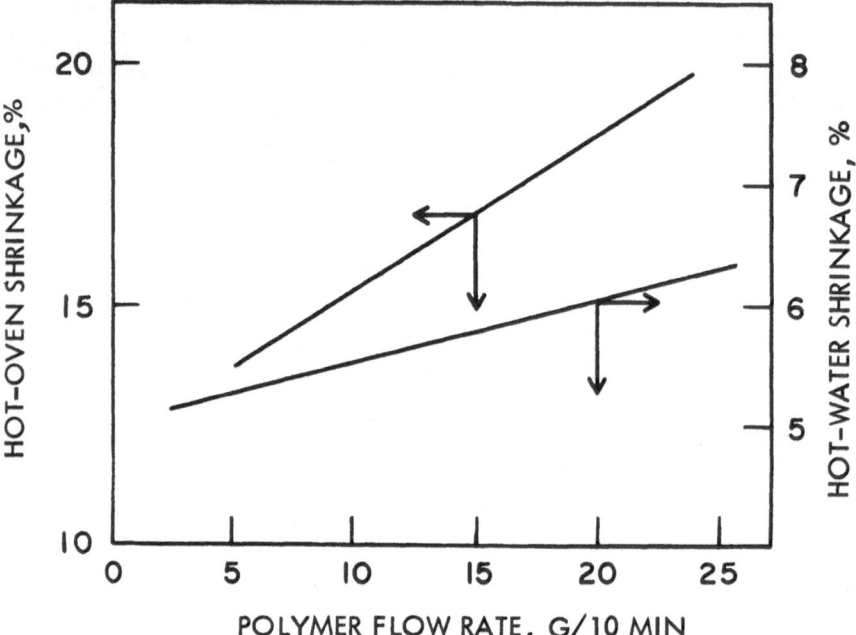

Figure 17. Effect of Polymer Flow Rate on the Hot-Air and Hot-Water Shrinkage of Polypropylene Monofilaments [11].

Additional effects of molecular weight on the tensile properties of isotactic polypropylene fibers have been reported by the Southern Research Institute; the fibers tested were prepared from Pro-fax polymers with the designation, molecular weight, and melt index as shown in Table 7. The relationships between molecular weight and polypropylene fiber properties as developed from this study are discussed below.

Figure 18 shows the effect of molecular weight on the variation of tenacity with total draw ratio. The tenacity increases slowly at low draw ratios (3 to 6), then increases sharply as the draw is raised from 7 to 15; further increase in maximum draw ratio increases the tenacity only slightly. In addition, a family of similar curves is generated which shows that the tenacity increases as the molecular weight of the filaments is increased from 67,000 to 405,000.

The variation of rupture elongation, modulus, and density with draw ratio for different molecular-weight polypropylene fibers is shown in Figure 19 as plots (a), (b), and (c), respectively; the draw ratio, modulus, and rupture elongation are plotted on logarithm coordinates while the density is shown on a Cartesian scale. Plots (a) and (b) show linear relationships between rupture elongation and modulus with draw ratio that are independent of fiber molecular weight; fiber density is independent of draw ratio up to a value of 15–20; higher draw ratios decrease the density to a minimum of about 0.8700 g/cc at a draw ratio of 50:1. Similar to curves (a) and (b), plot (c) shows that density is also independent of fiber molecular weight. Sheehan and Cole attribute this density decrease to (1) an attendant decrease in crystallinity with draw ratio as shown earlier in Figure 8, and (2) the formation of voids in the fiber upon drawing as reported by Wyckoff.

Figure 20 contains two plots which show the relationships among tenacity, modulus, and rupture elongation for fibers with molecular-weight ranges of 67,000–110,000 and 225,000–405,000. The rupture elongation decreases and the modulus increases as the fiber tenacity is increased. The effect of molecular weight on these fiber properties is manifested by an increase in rupture elongation and a decrease in modulus

TABLE 7

PROPERTIES OF ISOTACTIC POLYPROPYLENES TESTED BY THE SOUTHERN RESEARCH INSTITUTE

Polymer	Polymer molecular weight, $\overline{M}_w$		Melt index, g/10 min
	Determined by light scattering[a]	Determined in tetralin[b]	
6323	250,000	245,000	9.37
6423	290,000	285,000	7.89
6523	305,000	305,000	3.58
6623	335,000	315,000	2.71
6723	355,000	470,000	0.61

[a] Molecular weights as reported by the Hercules Powder Company, manufacturer.
[b] Molecular weights as measured by the Southern Research Institute.

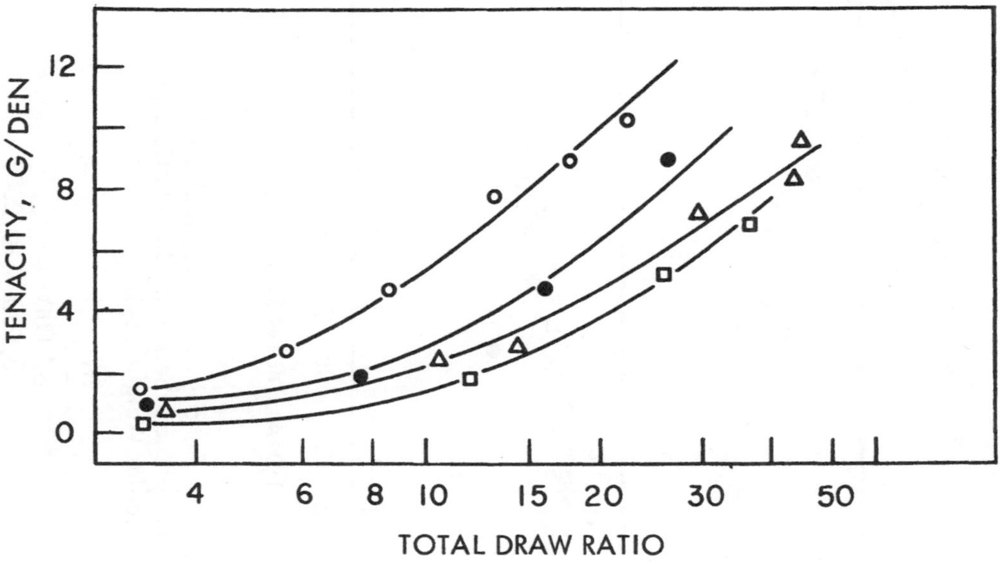

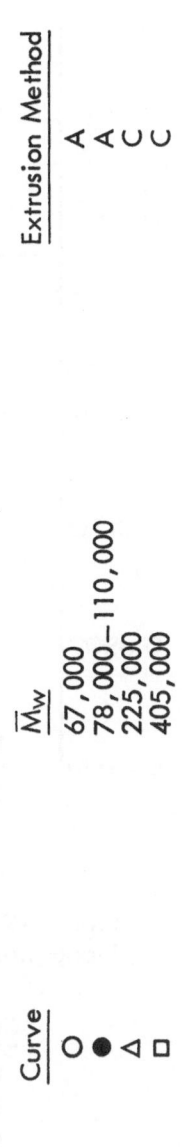

Figure 18. Effect of Draw Ratio and Molecular Weight upon the Tenacity of Polypropylene Fibers [34]:

Curve	$\overline{M}_w$	Extrusion Method
O	67,000	A
●	78,000–110,000	A
△	225,000	C
□	405,000	C

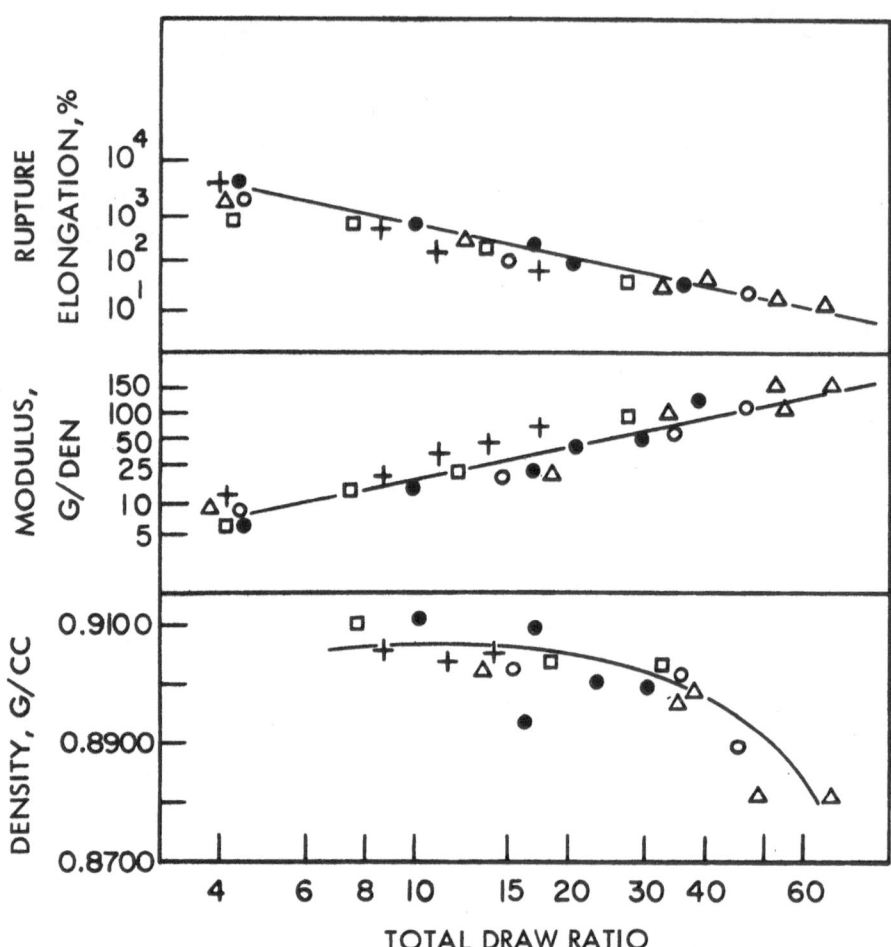

Figure 19. Variation of Density, Modulus, and Rupture Elongation of Polypropylene Fibers with Draw Ratio and Molecular Weight [34]:

Curve	$\overline{M}_w$
O	67,000
△	78,000–110,000
●	225,000
□	306,000
+	407,000

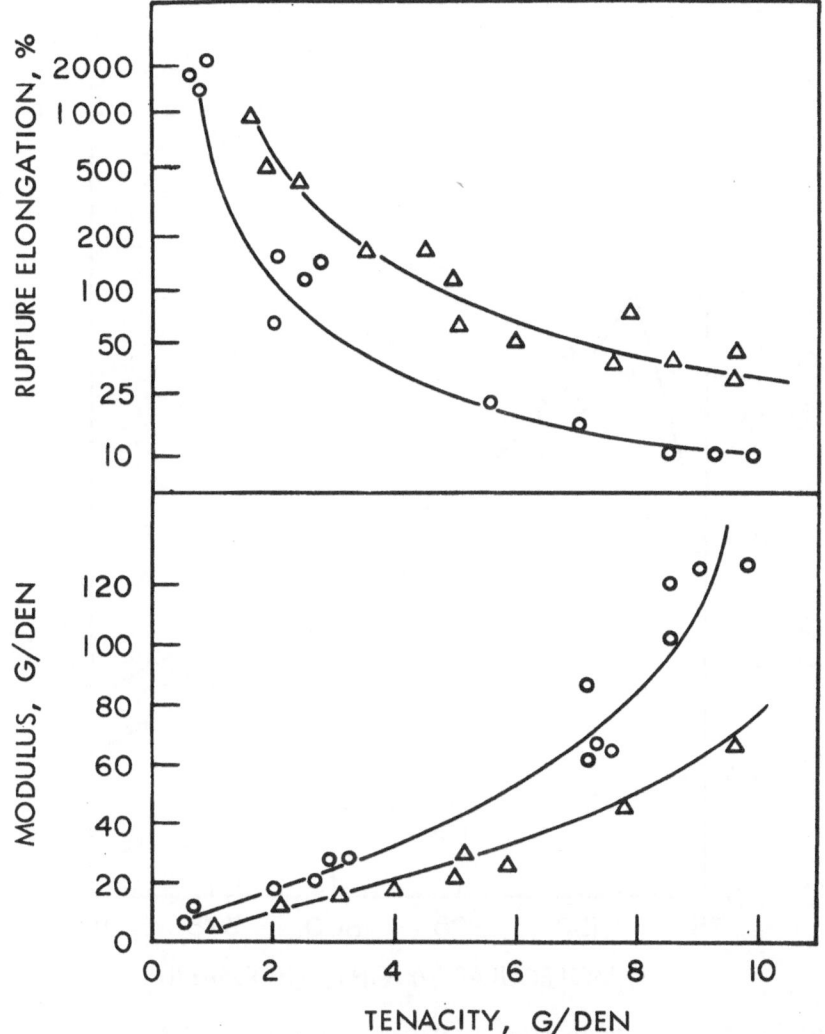

Figure 20. Relationship Between the Physical Properties of
Polypropylene Fibers and Molecular Weight [34]:

Curve	$\overline{M}_W$
O	67,000–110,000
△	225,000–405,000

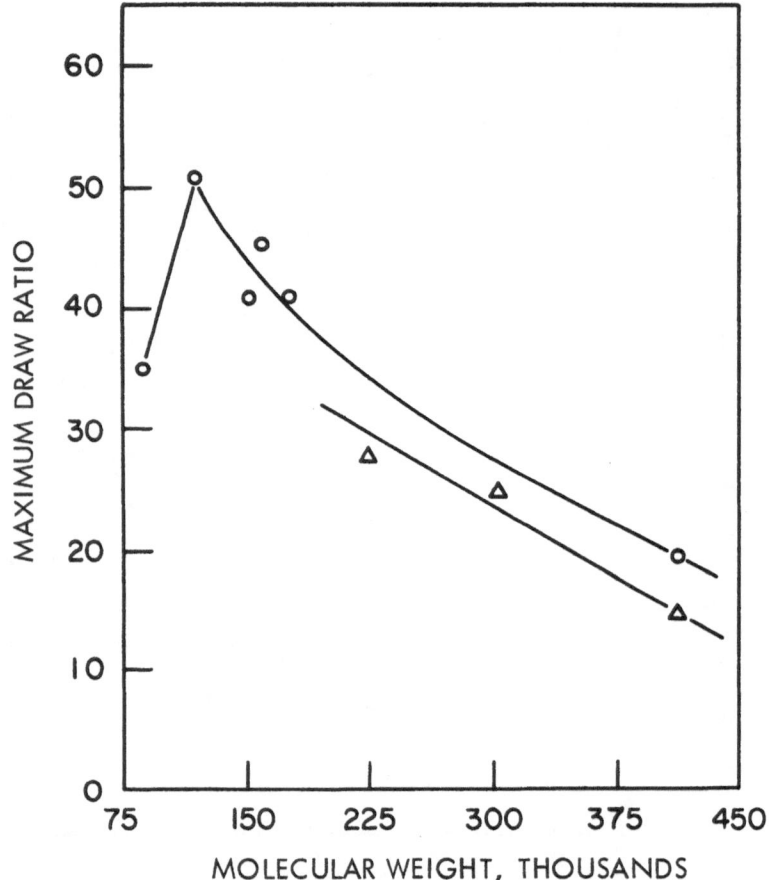

Figure 21. Maximum Draw Ratio of Crystalline and Amorphous Structures of Polypropylene Fibers as a Function of Molecular Weight [34]: (O) Amorphous. (Δ) Crystalline.

(at a specified tenacity value) as the molecular weight is increased from the lower to the upper molecular-weight range.

Figure 21 presents the maximum draw ratio of both the crystalline (monoclinic) and amorphous smetic structures of polypropylene fibers as a function of fiber molecular weight. The amorphous form exhibits a sharp increase in permissible draw ratio from 35:1 to 50:1 with a very slight increase in molecular weight from 75,000 to 140,000; the maximum draw ratio decreases rapidly as the molecular weight is increased further, reaching a value of about 15:1 at a molecular weight of 400,000. Though similar to the amorphous curve, the maximum draw ratio of the crystalline form is relatively lower for a given molecular-weight value. This plot illustrates the ease of drawing the amorphous structure compared to the monoclinic form as discussed previously; this increased drawing in turn produces high-tenacity polypropylene fibers.

3. RELATIONSHIPS BETWEEN FIBER FORMATION CONDITIONS AND POLYPROPYLENE FIBER PROPERTIES

a. Extrusion Conditions

(1) Effect of Polymer Viscosity Characteristics on Fiber Uniformity. The extrusion thermal requirements for any thermoplastic material are dictated by the flow behavior or viscosity characteristics of the polymer. This relationship of viscosity as a function of temperature is shown in Figure 22 for both polypropylene and polyethylene terephthalate at a common velocity gradient of $1 \cdot 10^{-2} \, \text{sec}^{-1}$. The curves show that high temperatures are required to obtain a fluidity in polypropylene comparable to that of polyethylene terephthalate; even at temperatures above the melting point, the viscosity of polypropylene remains at relatively high values. The effect of viscosity on polypropylene fiber properties is manifested by the uniformity of the extruded fiber; two undesirable phenomena of fiber uniformity related to polymer viscosity characteristics are swelling and irregularities, which are discussed below.

Swelling. In general, the jet of a polymer melt emerging from a capillary (such as fiber spinneret) swells to a diameter

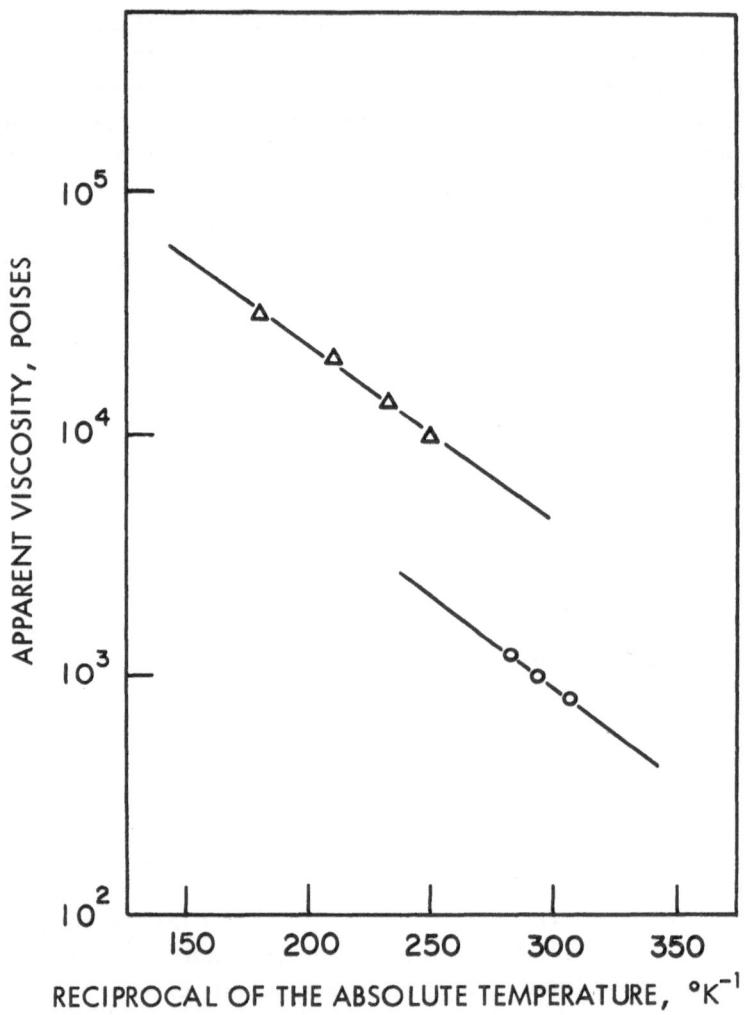

Figure 22. Variation of Viscosity as a Function of Temperature
for Polypropylene and Polyethylene Terephthalate [3]: (O) Polyethylene Terephthalate. (Δ) Polypropylene.

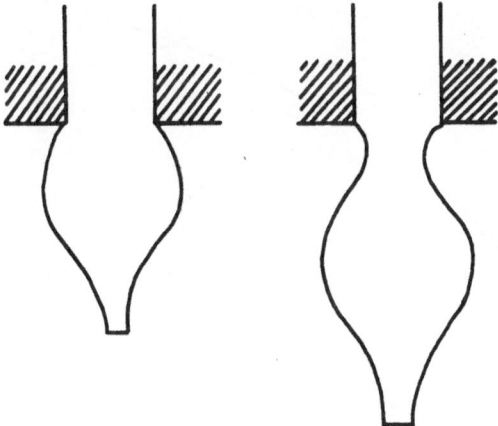

Figure 23. Characteristic Swelling of Molten Polypropylene upon Extrusion from a Spinneret in the Formation of Fibers [3].

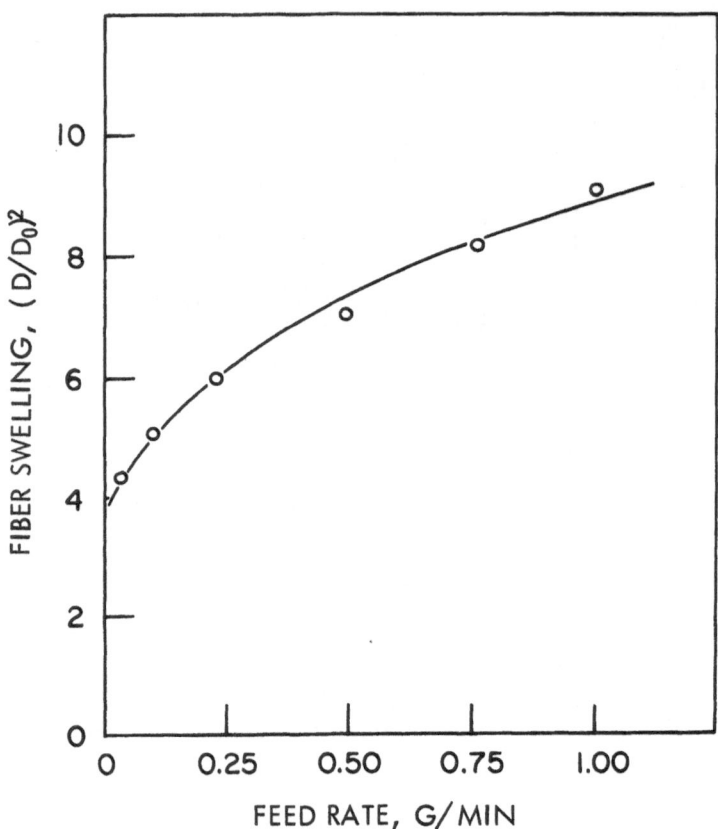

Figure 24. Relationship of Polypropylene Fiber Swelling,
$(D/D_0)^2$, to Polymer Feed Rate [3].

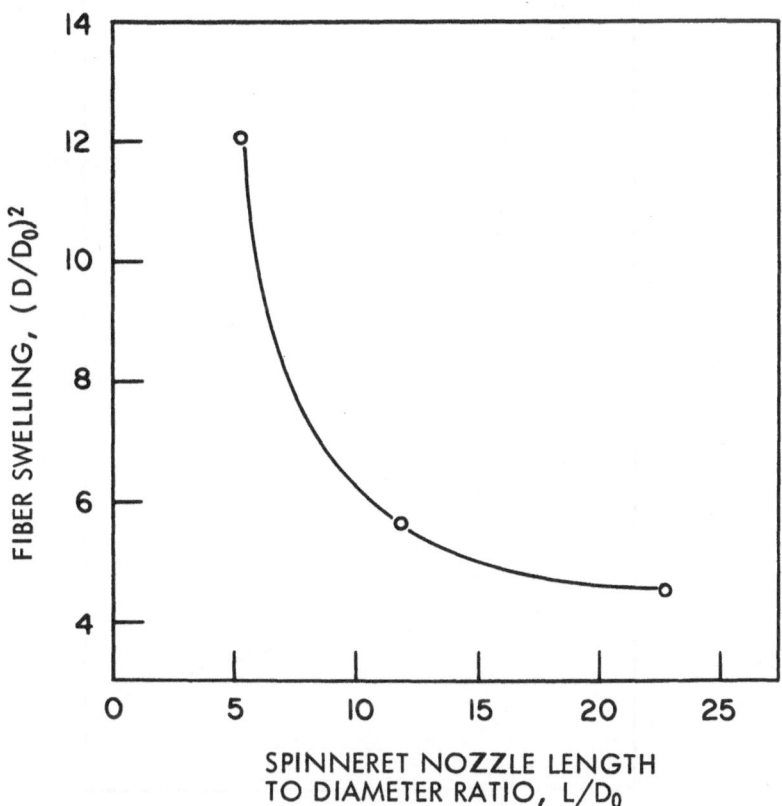

Figure 25. Relationship of Polypropylene Fiber Swelling, $(D/D_0)^2$, to the Ratio Between the Length and the Diameter of the Nozzle, L/D_0 [3].

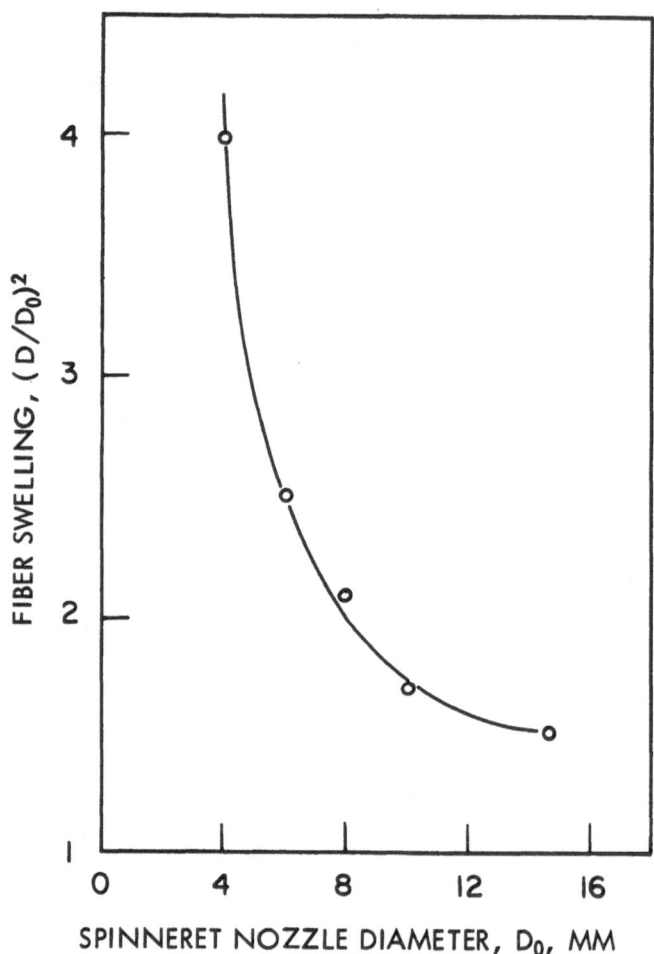

Figure 26. Relationship of Polypropylene Fiber Swelling, $(D/D_0)^2$, to the Diameter of the Spinneret Nozzle [3]: Feed Conditions: Temperature — 280°C. Feed Rate — 0.3 g/min. Ratio of Nozzle Length to Diameter — 20.

TABLE 8

INFLUENCE OF TEMPERATURE AND VISCOSITY ON
POLYPROPYLENE FIBER SWELLING $(D/D_o)^2$

Intrinsic viscosity, dl/g	Temperature, °C		
	190	230	280
1.27	2.8	1.5	1.3
1.90	4.2	2.8	2.1

Polymer feed rate. . . . . 0.5 g/min

Spinneret nozzle diameter. . . . . .0.8 mm

Length of the spinneret nozzle. . . . . 16 mm

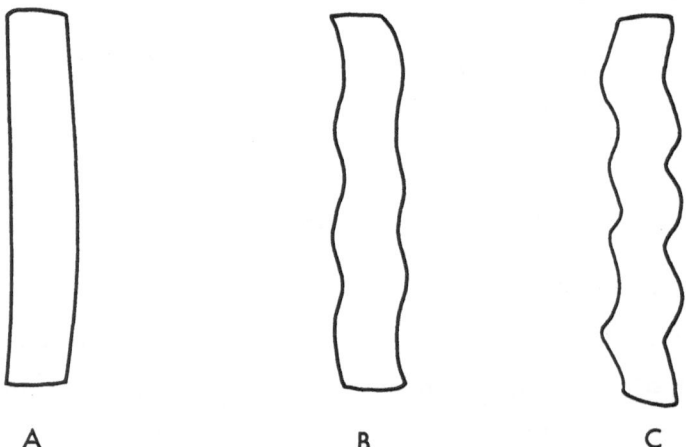

Figure 27.   Influence of Polymer Shear Stress on Polypropylene
Fiber Uniformity During Extrusion [3]: Relative Shear Stress:
(A) Normal. (B) High. (C) Excessive.

larger than the hole diameter. Polypropylene undergoes such swelling to a considerable extent even when the fiber is stressed, as during the stretching stage of manufacture. The profile of this swelling is shown in Figure 23. In addition to extrusion temperature and polymer characteristics, Cappuccio and others have shown that this swelling is influenced by the feed rate, spinneret nozzle diameter, D_0, and the ratio of the length to the diameter of the nozzle, L/D_0. These relationships are shown in Figures 24, 25, and 26, where swelling has been defined as the square of the ratio of the maximum diameter of the undrawn filament, D, and the nozzle diameter, D_0. As may be expected, polymer fiber swelling increases with melt viscosity and decreases at higher melt temperatures. Table 8 presents quantitative data on the effects of temperature and viscosity on swelling of polypropylene fiber; polymer feed rate and nozzle dimensions were unchanged during the tests. The practical aspect of this swelling phenomenon is found in the difficulty in attaining uniformity during the manufacture of very fine fibers.

Irregularities. Compared to the swelling characteristic discussed above, fiber uniformity is influenced more by another phenomenon which occurs at flow rates or shearing stresses that exceed certain critical limits. This latter occurrence, sometimes referred to as "melt fracture," results in fiber irregularities which are manifested by a transformation of the extruded filament from a cylindrical to a spiral shape, the degree of deformation depending on the applied stress. Figure 27 illustrates these fiber irregularities. It is reported that within a temperature range of 200–300°C and a polymer intrinsic viscosity of 1–3 g/dl, polypropylene exhibits these irregularities over the shear stress range of 0.8 to $1.2 \cdot 10^6$ dynes/cm^2.

From the above analysis, it is seen that fiber deformation occurring as the result of melt fracture is minimized by higher melt temperatures, lower polymer intrinsic viscosity, and increased nozzle diameter on the spinneret. Although these irregularities impose limits on the maximum permissible flow rates, alteration of the above processing conditions offer means

TABLE 9

EFFECT OF EXTRUSION CONDITIONS ON THE MOLECULAR WEIGHT AND TENACITY OF POLYPROPYLENE FIBERS

| Polymer | Extrusion method | Molecular weight, $\bar{M}_w$[a] | | Tenacity, g/den |
		Polymer	Filament	
6323	A	245,000	67,000	3.2
	C		225,000	4.8
6423	A	285,000	78,000	—
6523	A	305,000	88,000	—
6623	A	315,000	92,000	3.4
	C		305,000	6.3
6723	A	470,000	110,000	3.5
	C		405,000	8.1

[a]The molecular weights were determined from intrinsic viscosities of the materials in decalin.

of raising these limits. For example, since flow rate is proportional to the cube of nozzle diameter, this variable is most often selected when higher outputs are required.

(2) Degradation Effects of Air and Heat on Polypropylene Fiber Properties. The influence of extrusion conditions upon the physical properties of polypropylene filament has recently been investigated by the Southern Research Institute; additional, more limited data on the effects of process conditions on the properties of polypropylene have been developed by the Eastman Kodak Company. The background information on the type polymers investigated by the Southern Research Institute is presented in Table 7. In the Eastman Kodak study, the polymer tested was a Tenite polypropylene #4221 which is a high-molecular-weight material (about 340,000) with a flow rate of 2.5 g/10 min at 230°C and 2.16 kg load.

The combined degradation effects of air and temperature on molten polypropylene have been assessed by molecular-weight and tenacity measurements on fibers extruded under the following conditions: (1) Method A, in which no attempt was made to remove ambient air; extrusion temperature was 280°C. (2) Method C, in which the polymer and spinning pot were evacuated prior to extrusion at 235°C.

The data of the above tests are shown tabulated in Table 9 and graphically in Figure 28; the latter is a plot of the variation of tenacity with molecular weight for fibers produced under methods A and C described above. Table 9 shows that during the extrusion of type A fibers, polymer degradation results which causes a substantial decrease in molecular weight. Figure 28 shows that this molecular-weight reduction is most detrimental in the preparation of high-tenacity polypropylene fibers; the low-molecular-weight fibers exhibit a maximum tenacity of 4–5 g/den, while the tenacity for the high-molecular-weight fibers exceeds 9 g/den.

Figure 29 presents data from the Eastman Kodak study which show the effect of polymer melt temperature upon tenacity, elongation, and heat-oven shrinkage at constant draw ratio (8.8:1) and temperature (300°F). The plots show that shrinkage and elongation are not dependent on melt temperature,

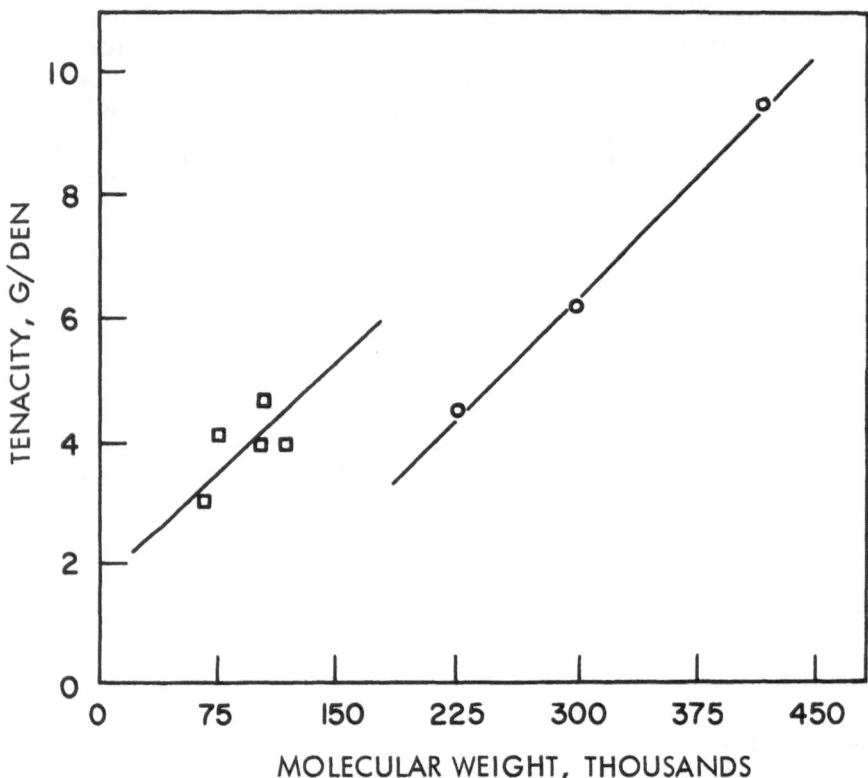

Figure 28. Relationship Between Tenacity and Molecular
Weight of Polypropylene Fibers [34]: (O) Extrusion Method C.
(□) Extrusion Method A.

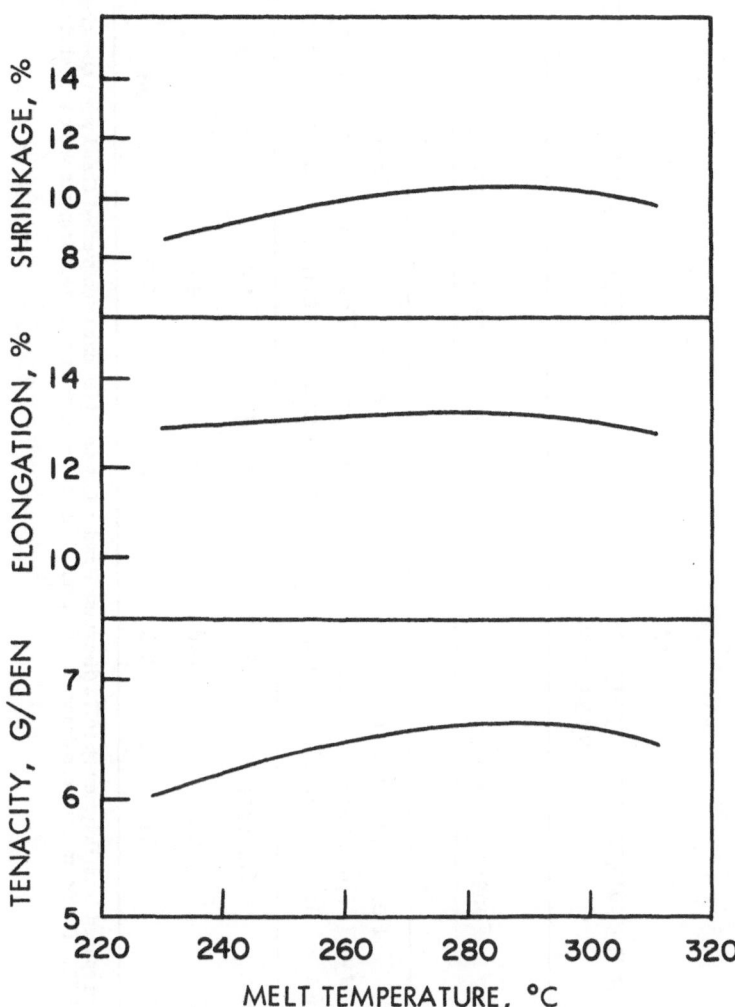

Figure 29. Effect of Melt Temperature on Polypropylene Fiber
Properties at a Draw Temperature of 150°C and a Draw Ratio
of 8.8:1 [28].

TABLE 10

RELATIONSHIP BETWEEN TENACITY OF POLYPROPYLENE FIBERS AND AMOUNT OF DRAW DURING SPINNING[a]

| | Tenacity and percent decrease in tenacity at total draw ratios of | | | | | |
| | 10 | | 15 | | 20 | |
Spin-draw ratio	Tenacity, g/den	Percent decrease in tenacity from control	Tenacity, g/den	Percent decrease in tenacity from control	Tenacity, g/den	Percent decrease in tenacity from control
3.1	4.9	Control	7.7	Control	9.3	Control
6.2	2.3	53	3.8	51	4.9	47
9.3	1.9	61	3.0	61	3.9	58

[a] The filaments were prepared from Polymer No. 6723 (Hercules Powder Company) by extrusion method C and were quenched in a water bath at 50°C. The filaments were drawn in a glycerine bath at 135°C and the drawn filaments were annealed for 15 min at 100°C.

TABLE 11

TENSILE PROPERTIES OF POLYPROPYLENE FILAMENTS PREPARED BY EXTRUSION METHODS A AND C AND QUENCHED UNDER DIFFERENT CONDITIONS[a]

Quenching conditions	Tenacity of drawn filaments, g/den
Extrusion method A, melt temperature, 280°C	
Water bath, 50°C.	6.5
Air, 20°C.	4.8
Extrusion method C, melt temperature, 235°C	
Water bath, 10°C.	9.5
Air, 20°C.	5.5

[a] The above fibers were prepared from Polymer No. 6723 (molecular weight of 350,000 – 470,000) from the Hercules Powder Company.

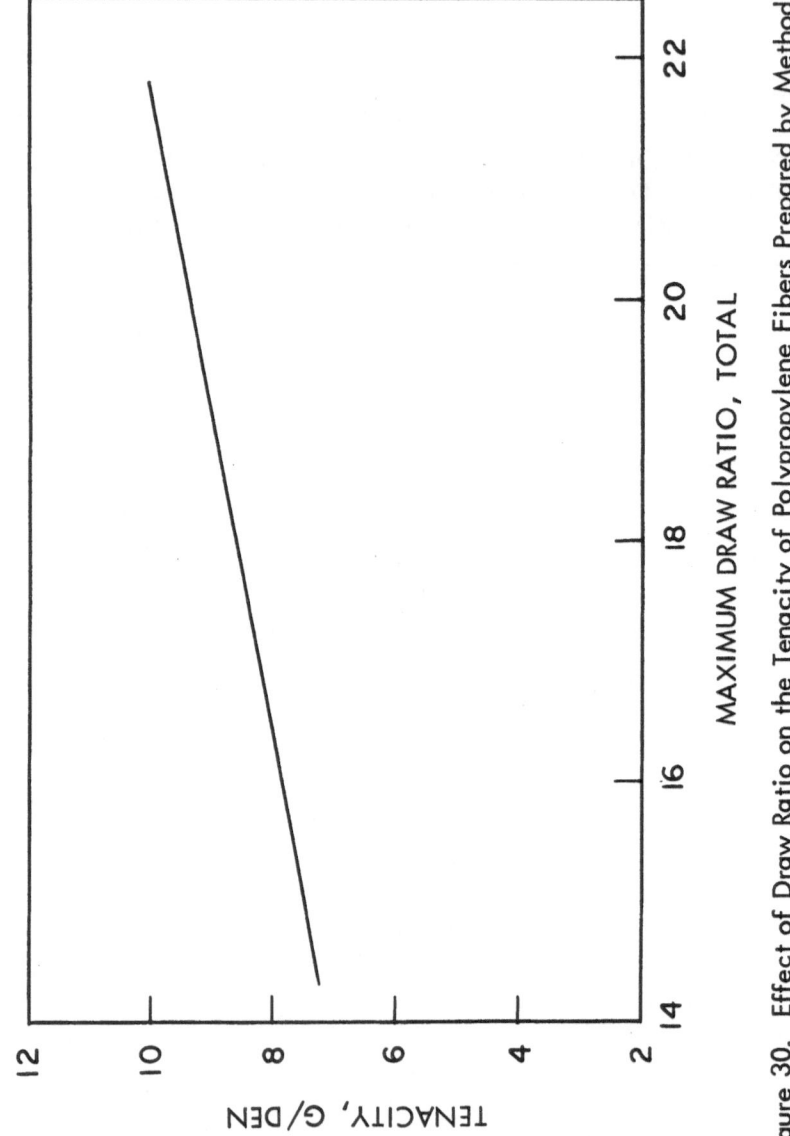

Figure 30.   Effect of Draw Ratio on the Tenacity of Polypropylene Fibers Prepared by Method C —
Air Evacuation Prior to Extrusion [30].

while tenacity is slightly adversely affected as the melt temperature is increased above 280°C.

The above data show that the serious degradation of polypropylene must be attributed to the combined action of oxidizing air and polymer melt temperature, or conversely, low melt temperature alone is insufficient to ensure high-tenacity polypropylene fibers.

b. Filament Draw During Spinning

Based on earlier discussions on the beneficial effect of high oven draw ratios on tenacity, one might anticipate even higher-tenacity fibers upon drawing the monofilament in the molten state, i.e., prior to quenching. Actually the converse of this relationship between polypropylene fiber tenacity and spin draw occurs, as indicated by the data in Table 10. The table shows a decrease in fiber tenacity up to 61% as the spin-draw ratio is increased from 3.1 to 9.3 for a given total draw ratio. The effects of various total draw ratios and spin-draw ratios upon polypropylene fiber tenacity are shown in the table.

c. Quench Medium and Temperature

The effects of quench medium and temperature on tenacity for method A- and method C- prepared fibers are shown in Table 11. The detrimental effect of the slow quench bath is attributed by Sheehan to a partial drawing of the fibers during spinning which, as discussed in the preceding section, decreases fiber tenacity significantly.

d. Oven Draw Ratio

Figure 30 shows the increase in fiber tenacity with increased draw ratio for annealed fibers prepared by spinning method C; the maximum total draw ratio obtained in this study was only 22:1, which corresponded to a tenacity of about 10 g/den. Higher-tenacity fibers were prepared by drawing fibers conditioned in an oven at 130°–135°C at draw ratios up to 34:1. Data obtained during these high draw ratio tests on tenacity, rupture elongation, and elastic modulus are presented in Table 12; at a draw ratio of 34:1, polypropylene fibers with tenacity values up to 13 g/den were prepared by the Southern

TABLE 12

PHYSICAL PROPERTIES OF POLYPROPYLENE FILAMENTS
DRAWN DIFFERENT AMOUNTS IN AN OVEN AT 130° — 135°C[a]

Draw ratio, total	Properties of drawn fibers			
	Denier	Tenacity, g/den	Elongation at break, %	Modulus, g/den
29	20.6	11.5	23	88
32	19.1	11.5	24	94
33	17.6	12.4	17	106
34	18.1	13.1	18	110

[a] The filaments were prepared from Polymer No. 6723 (Hercules Powder Company) by spinning method C with a quench temperature of 10°C.

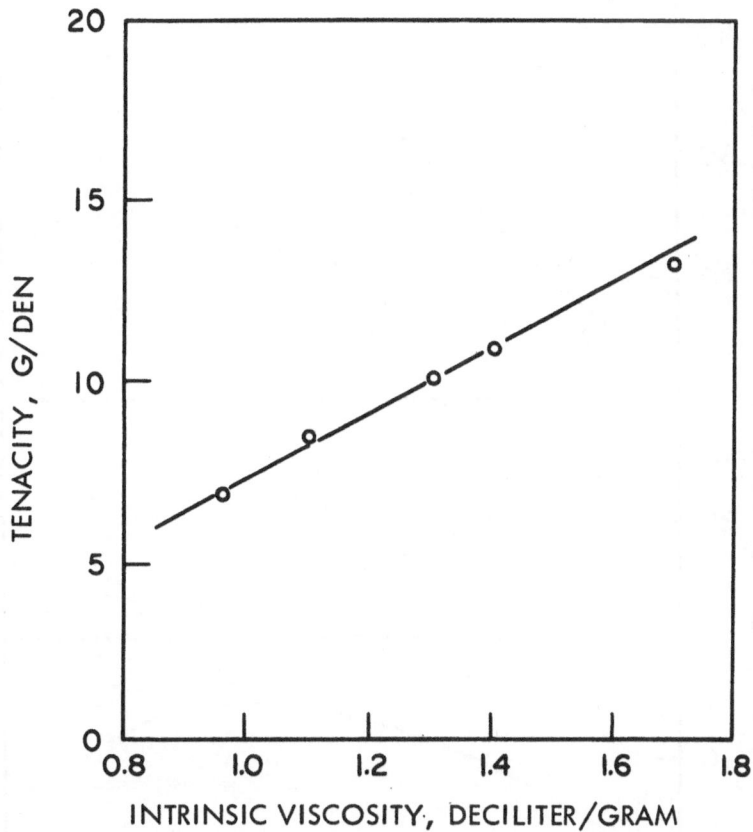

Figure 31. Tenacity of Polypropylene Fiber as a Function of the
Intrinsic Viscosity of the Polymer [3].

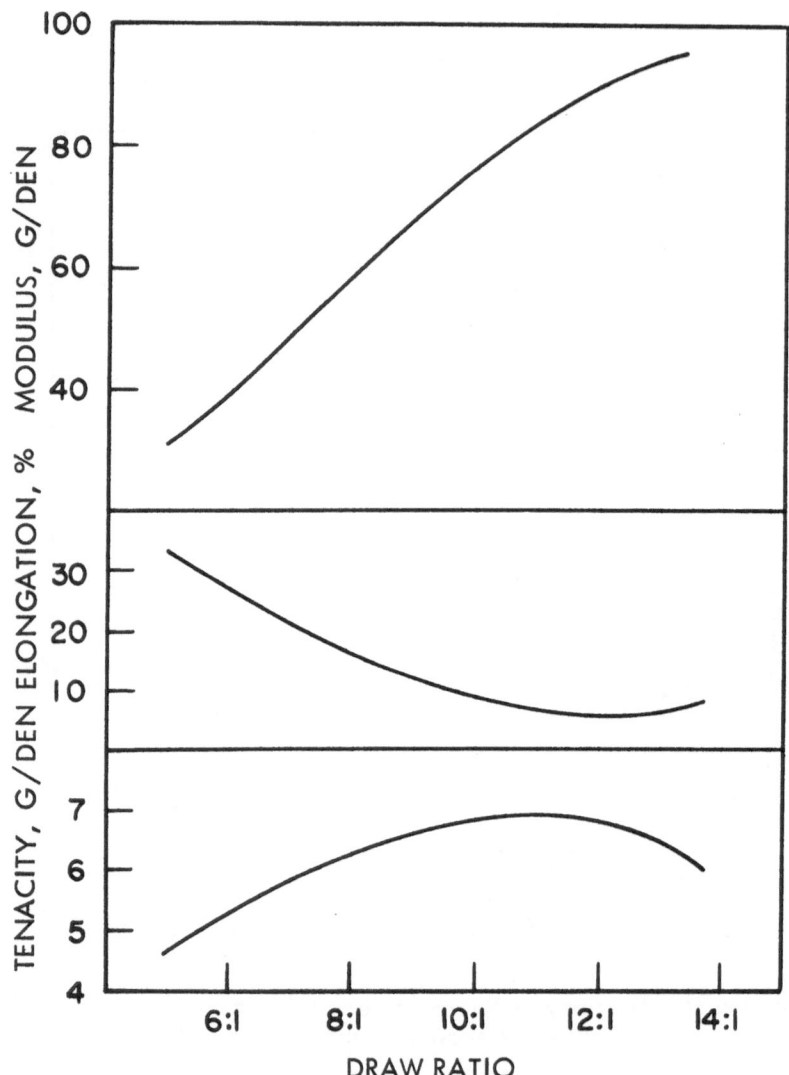

Figure 32.   Effect of Draw Ratio on Polypropylene Fiber Properties at a Draw Temperature of 300°F [28].

Research Institute.   In a similar work, Cappuccio et al. prepared polypropylene fibers with a tenacity up to 15 g/den by cold-drawing the amorphous form of polypropylene at high speeds.   This effect on tenacity is shown as a function of intrinsic viscosity in Figure 31.

Information on the variation of tenacity, rupture elongation, and elastic modulus at lower draw ratios, as developed by Eastman Kodak, is shown in the plots of Figure 32; the relative variation of the modulus and rupture elongation of the two investigations is similar; however, the tenacity reacts quite differently at the ranges of draw ratios studied. The lower draw ratios, up to 13, produce a maximum in tenacity of about 7 g/den, then the tensile strength decreases slowly as the draw ratio is increased further.   At the higher draw ratios of 29–34 (Southern Research Institute), the tenacity consistently increases from 11.5 to 13.1 g/den.

The above information shows that tenacities of polypropylene fibers commercially available usually range from 6–8 g/den, depending on the manufacturer.   Research works have indicated that much higher fiber tenacities — up to 15 g/den — can be prepared if desired.

e. Oven Draw Temperature

The effects of draw temperature, varied from 260°–380°F, upon fiber shrinkage, elongation, and tenacity are shown in Figure 33; a constant draw ratio of 8.8:1 was maintained during the tests.   The tenacity decreases slightly from 6.5 to 5.5 g/den with increasing draw temperature up to about 340°F; above this temperature no further change in tenacity occurs up to the maximum test value of 380°F.   A slight increase in elongation from 15 to 20% results in increasing draw temperature over the range tested.   The hot-oven shrinkage of the fiber decreases uniformly from 15% to less than 5% as the draw temperature increases up to 380°F.

The Southern Research Institute investigated the interrelationship between fiber draw and oven draw temperature on the properties of polypropylene fibers.   As discussed earlier, high-tenacity fibers are obtained by oven-annealing the fibers

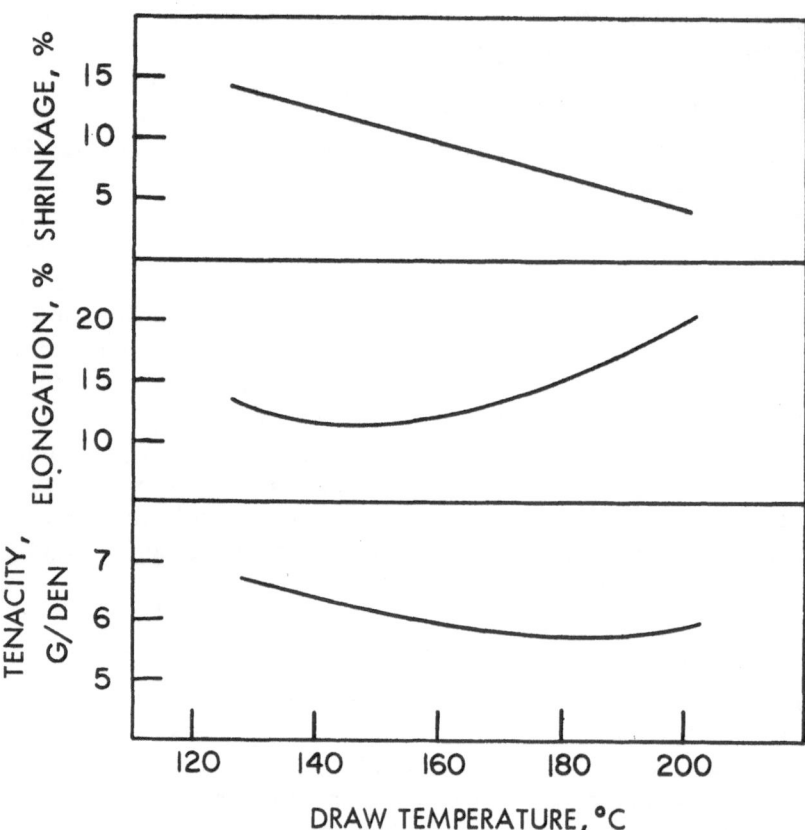

Figure 33. Effect of Draw Temperature on the Hot-Oven Shrink-age, Elongation, and Tenacity of Polypropylene Monofilaments [28].

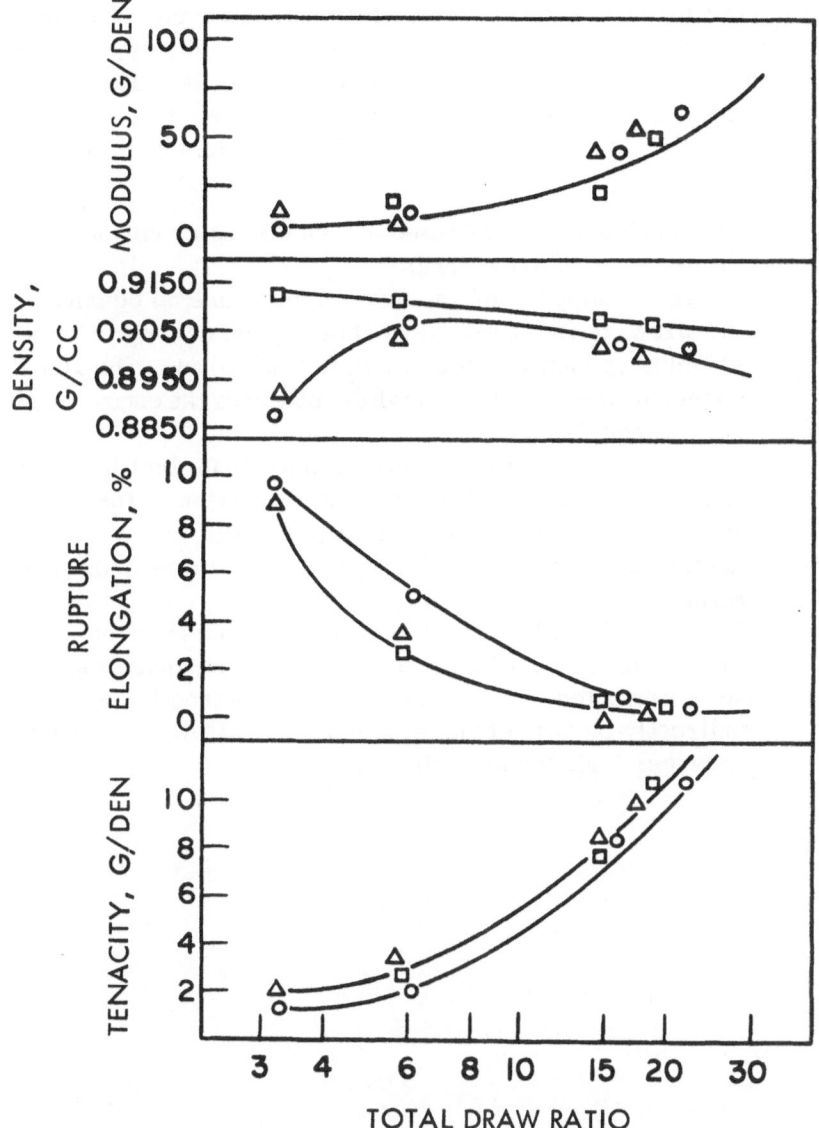

Figure 34. Relationship Between the Physical Properties of Polypropylene Fibers and Annealing Temperature [34]: (O) Unannealed. (Δ) Annealed at 100°C. (□) Annealed at 160°C.

to permit high draw ratios. The effect of these oven conditions, in relation to draw ratios, on the modulus, density, rupture elongation, and tenacity is shown in Figure 34. The figure contains data on fibers that were (a) unannealed, (b), annealed at 100°C, and (c) annealed at 160°C. The plots of Figure 34 show that:

1. The modulus is unaffected by annealing conditions, but increases with draw ratio.
2. Undrawn samples of (a) and (b) increase in density as the draw ratio is raised to about 7; the density of these samples is unaffected by further fiber draw. The 160°C annealed fibers exhibit a high density over the entire draw ratio range.
3. Annealing lowers the rupture elongation for both undrawn and drawn fibers. The percent elongation of the unannealed sample is slightly higher for the undrawn filaments but coincides with fibers (b) and (c) at high draw ratios.
4. Both annealed fibers show comparable effects on tenacity. The significance of this plot is that oven annealing per se contributes very little to fiber strength, but acts indirectly by permitting a high draw ratio which, in turn, produces high-tenacity fibers.

IV. Comparative Fiber Physical Properties

The effects of polymer characteristics and processing conditions as described in the preceding section are evidence of the various polypropylene fibers which can be produced with a wide range of properties. By proper control of the factors mentioned previously, a polypropylene fiber with optimum properties for a particular application can be specifically prepared. Such versatility in the properties of polypropylene fiber accounts for variances in the physical property data as reported by commercial manufacturers. The following information presented in this section on the physical properties of fibers has been generated by various investigators, each reporting on the physical behavior of a particularly prepared fiber. Although slight discrepancies may be noted, the data are adequate to permit a comparative analysis of the physical properties among the various fibers.

A. GENERAL PROPERTIES

1. SPECIFIC GRAVITY

The specific gravity of fibers determines the weight of the corresponding finished goods; from the standpoints of comfort (clothes) and ease of handling (tow ropes), a strong, lightweight fiber is most desirable. Further, a lower-density fiber affords appreciable cost savings in the extra yield per pound of material. The fact that polypropylene enjoys these advantages to the maximum relative to currently available

TABLE 13

RELATIVE COVERING POWER AND SPECIFIC GRAVITY OF
VARIOUS FIBERS

Fiber	Relative coverage power, lb	Specific gravity
Polypropylene	1.00	0.90–0.92
Cotton	1.71	1.50–1.55
Wool	1.46	1.30–1.32
Viscose rayon	1.70	1.52
Acetate	1.47	1.32
Nylon 6, 66	1.26	1.14
Orlon acrylic	1.30	1.14–1.17
Dacron	1.53	1.38–1.39
Dynel	1.44	1.30
Saran	1.88	1.72
Glass fiber	2.82	2.54

TABLE 14

THERMAL CONDUCTIVITY OF VARIOUS FIBERS
RELATIVE TO AIR

Material	Heat conductivity, relative to air
Air	1.0
Polypropylene	6.0
Wool	6.4
Acetate	8.6
Viscose	11.0
Cotton	17.0

TABLE 15

EFFECT OF DYES AND TEMPERATURE ON THE ELECTRICAL PROPERTIES OF POLYPROPYLENE MONOFILAMENTS

Monofilament	Volume resistivity at 30°C, Ω-cm	Volume resistivity at 130°C, Ω-cm	Dielectric constant at 1 kc and 25°C	Dielectric constant at 1 kc and 70°C	Power factor at 1 kc and 25°C	Power factor at 1 kc and 70°C
Normal	$5 \cdot 10^{15}$	$9 \cdot 10^{12}$	2.10	2.06	$3 \cdot 10^{-4}$	$3 \cdot 10^{-4}$
Red	$1 \cdot 10^{15}$	$8 \cdot 10^{12}$	2.10	2.04	$12 \cdot 10^{-4}$	$17 \cdot 10^{-4}$
Yellow	$1 \cdot 10^{15}$	$6 \cdot 10^{12}$	2.10	2.05	$6 \cdot 10^{-4}$	$8 \cdot 10^{-4}$
Green	$3 \cdot 10^{15}$	$1 \cdot 10^{12}$	2.10	2.05	$4 \cdot 10^{-4}$	$7 \cdot 10^{-4}$
Blue	$1 \cdot 10^{15}$	$0.5 \cdot 10^{12}$	2.10	2.03	$5 \cdot 10^{-4}$	$8 \cdot 10^{-4}$
Black	$2.5 \cdot 10^{15}$	$8 \cdot 10^{12}$	2.12	2.06	$4 \cdot 10^{-4}$	$1 \cdot 10^{-5}$

fibers is illustrated by Table 13, which lists comparative fiber specific gravity as well as the equivalent poundage of other fibers required to cover an area comparable to that covered by polypropylene. The significance of this lowest-specific-gravity fiber is realized by the greater surface and volume of finished products per unit weight obtained from polypropylene fibers.

2. THERMAL CONDUCTIVITY

Another outstanding property of polypropylene fiber is its thermal conductivity, which is the lowest of all fibers, a distinction previously held by wool. This characteristic makes polypropylene the best choice, relative to other available fibers, for insulating applications, such as cold-resistant clothes. Table 14 presents the heat conductivity of several materials relative to air.

3. ELECTRICAL PROPERTIES

The electrical characteristics of polypropylene are presented in Tables 15, 16, and 17. Table 15 shows the effect of dyes and temperature on the volume resistivity, dielectric constant, and power factor. Table 16 contains the variation of the dielectric constant and power factor with frequency. Table 17 compares the electrical properties of polypropylene with other thermoplastic materials at the standard conditions of 1 Mc, 64% relative humidity, and 23°C.

From Table 15, it is seen that as the temperature increases from 30° to 130°C, the volume resistivity of polypropylene decreases by a factor of about 10^3; colored fiber has little effect on the resistance. The dielectric constant increases with the addition of black pigment, and decreases with higher temperatures. The power factor of pigmented polypropylene fiber increases with higher temperatures and is essentially independent of the dyes except for red, which imparts a four-fold increase, relative to normal polypropylene at 25°C.

Table 16 shows that the power factor increases by a factor of three when the frequency is increased from 60 cps to 10 kcps;

TABLE 16

VARIATION OF THE DIELECTRIC CONSTANT AND POWER
FACTOR WITH FREQUENCY FOR NATURAL-COLORED
POLYPROPYLENE[a]

Frequency, cps	Dielectric constant	Power factor
60	2.26	$5 \cdot 10^{-4}$
$2 \cdot 10^2$	2.26	
10^4	2.26	$1.3 \cdot 10^{-3}$
10^5	2.25	
10^6	2.25	$5 \cdot 10^{-4}$

[a] Measured at 23°C and 50% relative humidity.

the power factor decreases to its original value upon further increase in frequency to 1 Mc. The dielectric constant undergoes no significant change with frequency.

The data in Table 17 show that polypropylene compares favorably in electrical characteristics with other competitive materials. In summary, polypropylene can be considered as a very good insulator with exceptionally low power loss even at radio frequencies.

B. VISCOELASTIC BEHAVIOR

1. GENERAL CONSIDERATIONS

The elongation of polymeric materials such as plastics, rubbers, and elastomers is not linearly proportional to an applied load, i.e., Hooke's law (strain or elongation is proportional to the stress or load) is not followed. These materials which do not conform to a simple linear relationship between load and elongation are termed as "viscoelastic" or non-Newtonian. Under load, viscoelastic materials manifest periods of perfect elasticity as well as permanent deformation. This characteristic becomes evident upon consideration of the stress–strain curves of these materials; with respect to fibers, the stress is usually expressed in terms of tenacity (grams/denier), and the strain in percent elongation. Data on several important fiber tensile properties can be obtained from the shape of these curves: rupture stress, rupture elongation, modulus of elasticity, yield strain, and toughness, the latter representing the energy required to rupture the fiber. A stress–strain diagram is constructed by measuring the elongation, which results in loading the fiber until rupture. The elastic response of fibers to strain depends on (1) temperature, (2) relative humidity, and (3) rate of elongation, or time during which the strain was applied. While the load–extension curves of the natural fibers, such cotton, wool and silk, are predetermined, the synthetic fibers can be prepared with a wide range of resistance to elongation, which is mainly dependent

TABLE 17

ELECTRICAL PROPERTIES OF VARIOUS
THERMOPLASTICS

Material	Dielectric constant at 1 Mc	Power factor at 1 Mc	Dielectric strength (V/mil)
Polypropylene	2.2	0.0005	500
Polyethylene	2.2	0.0005	500
Polystyrene	2.4	0.0001	500
Nylon	3.5	0.03	400
Saran	3.0	0.05	350
Teflon	2.0	0.003	500

on the degree of orientation of the fiber molecules. This orientation is controlled by the amount of stretch applied to the fiber during manufacture.

2. FIBER STRENGTH AND TENACITY

The breaking strength of fibers can be expressed, as with all engineering materials, in terms of load (pounds) per unit area (inches); however, because the cross sections of fibers vary widely and are irregular in shape, fiber strengths are generally based on strength per unit length. This stress for fibers is usually expressed as grams/denier, where denier is by definition the weight in grams of 9000 m of fiber.

The concept of tenacity (weight/length) compared to strength (weight/area) requires only the measurement of fiber length, which is much simpler than that of fiber cross-sectional area; the disadvantage of the use of tenacity is its dependence upon fiber specific gravity, whereas strength on a unit area basis is absolute. The relationship between strength and tenacity is

TENSILE STRENGTH (LB/IN.²) = 12,800 · SPECIFIC GRAVITY · TENACITY (G/DEN)

The following data show that in terms of fiber terminology nylon is stronger than steel, but on an absolute basis, steel can withstand loads five times greater than nylon:

Material	Tenacity, g/den	Tensile strength, psi
Nylon	9.0	100,000
Steel	3.5	500,000

Comparative fiber tenacities, both dry and wet values, are given Table 18. The wet tenacity is given in terms of percent of the dry tenacity. Fibers are generally converted into yarns; supplemental data on the tenacity of various yarns are presented in Table 19. The information from these tables shows that (1) the tenacities of fiber and yarn prepared from polypropylene are comparable to the more expensive nylon in the dry state and

TABLE 18

WET AND DRY TENACITIES FOR VARIOUS FIBERS

Fiber	Dry tenacity, g/den	Wet tenacity, % dry tenacity
Cotton, raw	3.0–4.9	100–110
Wool	1.0–1.7	76–97
Teflon fluorocarbon	1.7	100.0
Glass	6.0–7.3	65.0
Nylon 6, regular	4.5–5.8	91.3–95.5
Nylon 6, high tenacity	6.8–8.6	79.4–87.2
Nylon 66, regular	4.6–5.9	87–88
Nylon 66, high tenacity	5.9–9.2	86.4–85.8
Polyethylene, low density	0.5–2.0	100
Polyethylene, high density	4.5–8.0	100
Polypropylene	5.5–8.0	100
Polyvinyl alcohol	4.5–6.0	80.0–85.0
Dacron polyester, regular	4.4–5.0	100
Dacron polyester, high tenacity	6.3–7.8	100
Steel	3.5	100

TABLE 19

TENACITIES OF VARIOUS YARNS

Yarn	Tenacity, g/den
Polypropylene	8.0–8.5
Nylon	7.0–8.8
Polyester	6–7
Cotton	1.5–2.0
Viscose rayon	3.3–3.8
Polyethylene	4.5–6.0
Glass	6–7
Steel	3.5

TABLE 20

VARIATION OF TENACITY OF POLYETHYLENE AND
POLYPROPYLENE OVER A TEMPERATURE RANGE
OF 0°–60°C

Temperature, °C	Tenacity, g/den		
	Low–density polyethylene	High–density polyethylene	Polypropylene
0	1.7	8.4	8.0
20	1.2	6.1	6.5
40	0.95	4.0	5.4
60	0.6	2.5	5.1

TABLE 21

PERCENT STRENGTH LOSS OF VARIOUS FIBERS UPON
THERMAL EXPOSURE

Fiber	100°C	130°C	Exposure time, days
Polypropylene	0	20	5
Dacron	0	5	20
	4	25	80
Orlon	0	9	20
	0	45	80
Glass	0	0	20
	0	0	80

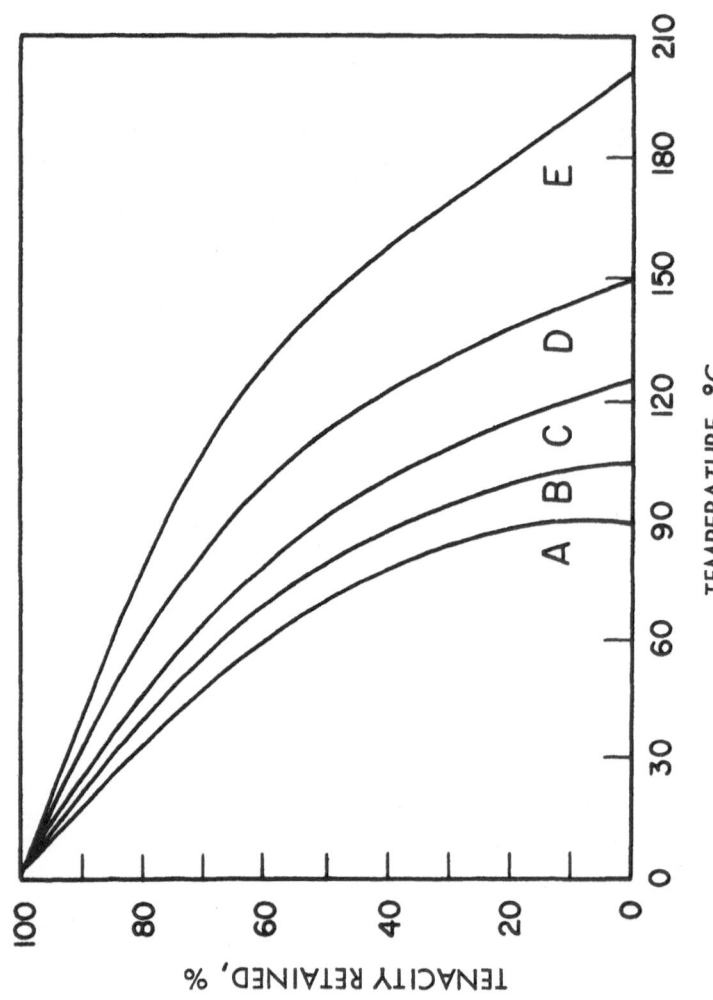

Figure 35. Effect of Elevated Temperatures on the Strength of Various Fibers [16]:
(A) Polyethylene, Low Molecular Weight. (B) Polyethylene, Medium Molecular Weight.
(C) Polyethylene, Linear. (D) Polypropylene. (E) Nylon 66.

(2) in the wet condition, polypropylene fiber and yarn are considerably stronger than any other textile material currently available. The wet and dry tenacities of polypropylene are identical because water does not permeate the fiber. The outstanding strength of polypropylene is further emphasized by the fact that fibers with tenacities exceeding 10 g/den have been tested experimentally as mentioned earlier in this report; such evidence accounts for the anticipation of a significant increase in the strength of commercial polypropylene fiber in the near future.

Table 20 contains the variation of tenacity of polypropylene and polyethylene (high- and low-density types for the temperature range of 0°–60°C. These data show that the decrease in strength with increasing temperature is much more severe for polyethylene (8.4 to 2.5 g/den) relative to polypropylene (8.0 to 5.1 g/den). Additional data are presented in Figure 35, with plots of the tensile strength expressed as tenacity retained upon thermal exposure for polypropylene, polyethylene, and nylon fibers. As would be expected from its higher melting point (250°C), the strength of nylon is less affected by thermal exposure than is the strength of polypropylene. Similarly, the heat resistance of dacron polyester and orlon also exceeds that of polypropylene, as illustrated by the heat-aging data of Table 21.

The increased thermal resistance of polypropylene fiber in the form of 840-den yarn is shown in Figure 36; the tenacity was measured after a 10-min exposure time. The interesting feature of this plot is the retention of tenacity at exposure temperatures above the melting point of the fiber (175°C).

3. ELASTIC MODULUS

The ratio of the change in stress to the change in strain within the elastic limits of a material is defined by ASTM as Young's modulus. However, in regard to synthetic fibers, the modulus is constant only for the initial portion of the linear load—elongation diagram as a result of the viscoelastic behavior of these organic materials. Consequently, the modulus

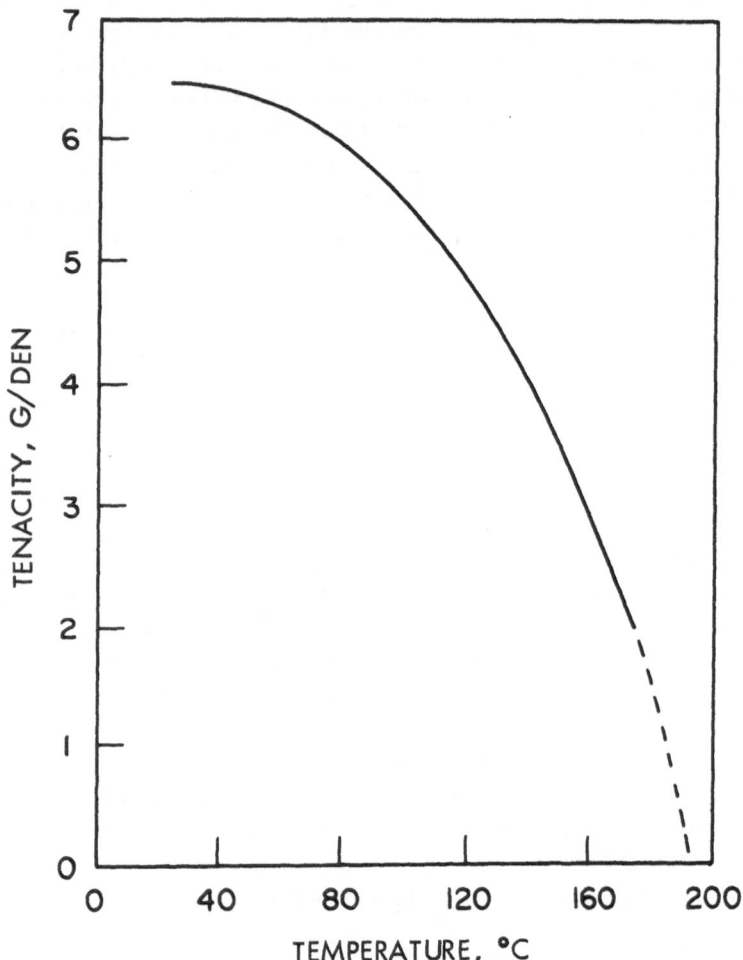

Figure 36. Strength Retention of Polypropylene Yarn
(840 den) as a Function of Temperature [21].

of fibers is based on an elongation at a rate equal to the initial slope of the stress—strain curve. The replacement of pounds per square inch with tenacity for units of stress, as required by fiber nomenclature, gives the modulus in terms of grams/ denier per unit strain or simply as grams/denier.

Comparative data on the initial modulus of fiber elasticity are presented in Table 22; these data are based on a 1% stretch, which is a strain small enough to ensure that the ratio of stress to strain is constant. The table shows that the most extensible fibers, such as wool and polyethylene, exhibit the lowest moduli in the range of 25–30 g/den; the brittle materials, such as polyesters, glass, and steel, have the highest moduli (about 300 g/den) because these fibers are unable to deform under load. With a modulus of 80–100 g/den, polypropylene is suitable for many applications because it is neither too stiff nor too elastic; nylon is considered quite extensible with a modulus about 45 g/den. The "ultimate" fiber is one that possesses both the temperature resistance of graphite, ceramic, or metallic substances and a modulus of about 100 g/den to permit sufficient deformation under load, i.e., is not brittle.

Whereas the elastic modulus of fibers is the ratio of load to elongation based on the initial (linear) portion of the stress— strain diagram, a similar property of fibers—average stiff- ness—is defined as the ratio of the breaking load and the elongation; the ASTM designation for average or mean stiff- ness is "secant modulus." Comparative data on mean stiff- ness (g/den) are presented in Table 23. The moderate elasti- city of polypropylene, 25–39 g/den stiffness, further emphasizes the relative strength of this material.

4. STRESS – STRAIN RELATIONSHIPS

a. Fiber Rupture Elongation

The percent elongation (wet and dry) to rupture various fibers under static conditions is listed in Table 23. In general, the strength and elongation of fibers are inversely proportional: Fibers such as glass and steel are very strong, but brittle, and

TABLE 22

INITIAL MODULUS OF ELASTICITY OF VARIOUS FIBERS

Fiber	Modulus, (grams per denier per unit strain)
Steel	280
Glass	307
Manila	250
Cotton	55
Flax	200
Wool	25–30
Rayon	75–175
Polyester	130
Nylon	45
High-density polyethylene	30
Polypropylene	80–100

TABLE 23

STIFFNESS, RUPTURE ELONGATION, AND TOUGHNESS OF FIBERS

Fiber	Breaking elongation		Average stiffness, g/den	Toughness index, g-cm/den-cm
	Dry, %	Wet, %		
Cotton, raw	3–7	–	60–70	0.15
Wool	25–35	25–50	4.5	0.35
Glass	3.4	2.5–3.5	177–215	0.10–0.12
Nylon 6, 66	20–40	20–40	15–45	0.5–0.7
Polyethylene, low density	20–80	20–80	6–10	0.20
Polyethylene, high density	10–20	10–20	22–80	0.40–0.45
Polypropylene	17–22	17–22	25–39	0.52–0.60
Dacron polyester, regular	14–25	19–25	18–36	0.35–0.55
Dacron polyester, high tenacity	10–14	10–14	45–78	0.31–0.42
Steel	8	8	44	0.14

TABLE 24

FIBER LOOP AND KNOT STRENGTHS

Fiber	Loop strength, g/den	Knot strength, g/den
Acrilan acrylic	—	2.0–2.3
Orlon acrylic	1.1–1.9	2.7–3.0
Acetate	0.49	1.0–1.2
Teflon fluorocarbon	1.2	11.7
Glass	1.0	16.2–19.8
Nylon 6, regular	3.8–5.4	34.2–48.6
Nylon 6, high tenacity	7.0	54.9
Polyethylene, high density	6.2–13.0	27.0–40.5
Polypropylene	4.5–5.2	36.0–40.5
Dacron polyester, regular	3.4–4.7	—
Dacron polyester, high tenacity	5.1	—
Viscose rayon, high tenacity	2.3–2.5	19.8–21.6

consequently do not yield under load, while the weak fibers, such as wool, extend easily. Tough and brittle fibers have breaking elongations in the ranges of 35–40 and 3–10, respectively, expressed as percentages. On a dry basis, the moderate extension prior to rupture classifies polypropylene as a tough fiber, compared to the nylons. With regard to breaking elongations in the wet state, moisture acts as an internal plasticizer which promotes fiber extensibility. Since most synthetic fibers are hydrophobic, the wet and dry breaking elongations are essentially identical, as exemplified by the polyolefins, polyesters, and acrylics; the notable exception is nylon, which, due to molecular polar groups, absorbs sufficient moisture to adversely affect the breaking elongation.

b. Toughness

The toughness of a fiber is a measure of the ability of the material to absorb energy. This energy absorption (or work done) is quantitatively determined by the area of the load—elongation curve in accordance with the following expression:

$$\text{TOUGHNESS} = \tfrac{1}{2} \left[\text{BREAKING LOAD (G/DEN)} \cdot \text{BREAKING ELONGATION (CM/CM)} \right]$$

Table 23 also presents relative toughness data for various fibers. The low elongation of the strong fibers—steel and glass—accounts for low energy absorption and the subsequent failure of these materials when stressed. The weak fibers, such as wool and branched polyethylene, are easily deformed and, consequently, can absorb much energy; however, the very low breaking load of these fibers is a serious handicap. The data show that polypropylene ranks well with the toughest fibers of cotton, nylon, and dacron polyester; these tough fibers have energy absorption values of about 0.5–0.7 g-cm/den-cm. The variation of fiber toughness as a function of strain rate is discussed in a later section.

c. Loop and Knot Strengths

In general, the physical properties of a fiber are essentially independent of the fiber diameter; two notable exceptions are fiber loop and knot strengths. The loop strength is the tensile load required to rupture two interlaced fiber

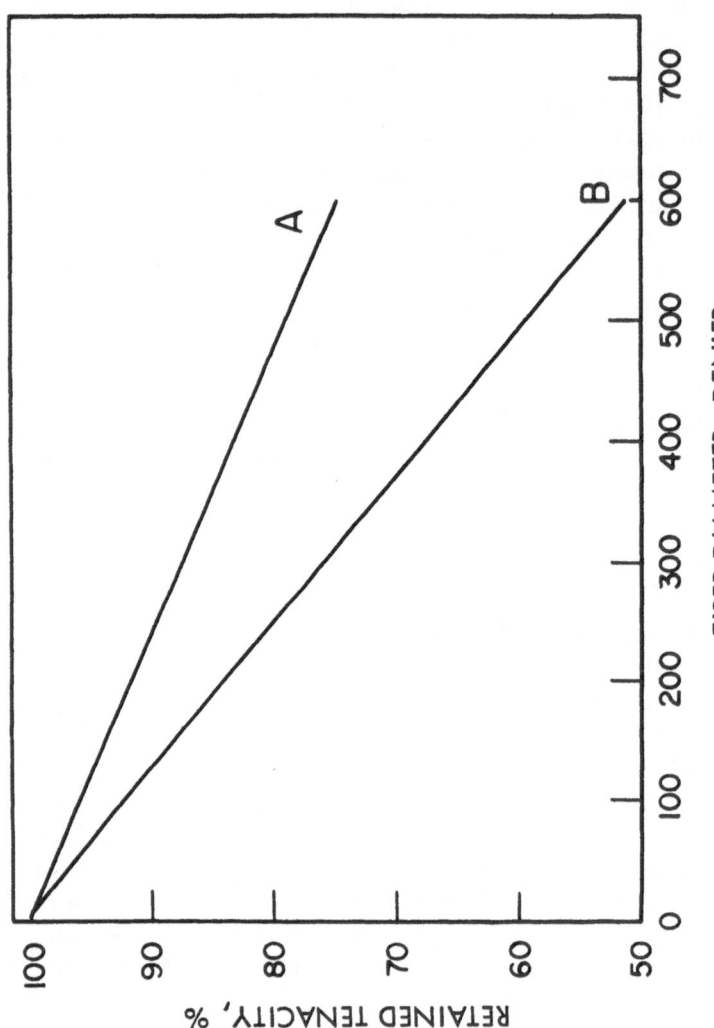

Figure 37. Effect of Polypropylene Fiber Diameter on Knot and Loop Strengths [5]: (A) Knot Strength. (B) Loop Strength.

loops; the looping resistance is the ratio of this breaking load of the looped fibers to twice the breaking load of the fiber alone multiplied by 100. The knot strength is the tensile load required to rupture a fiber looped into a simple overhand knot; the knotting resistance is the ratio between the breaking load of a fiber with the knot multiplied by 100 to the breaking load of the unknotted fiber.

Upon elongation, brittle fibers quickly develop concentrated stresses which cannot be dissipated and tensile failure results. Therefore, the brittle fibers, which have a high modulus of elasticity and high strength properties, possess poor loop and knot strengths. In contrast, more extensible fibers with relatively much higher rupture elongations than the brittle fibers are characterized by high loop and knot strengths; these latter properties result from the fact that the elastic fibers can stretch under stress, to an extent, rather than fail immediately.

The outstanding loop and knot strengths of polypropylene are shown by the comparative data of Table 24; only polyethylene and the more expensive nylon exceeds the loop strength of polypropylene, while with regard to knot strength, polypropylene ranks second only to nylon. The variation of polypropylene knot and loop strengths as a function of fiber diameter is given in Figure 37; the influence of temperature on the knot strength of Herculon monofilament is shown by Figure 38. These plots show that (1) the knot and loop strengths of polypropylene decrease sharply with increasing fiber diameter and (2) the knot strength of polypropylene increases as fiber temperature decreases.

d. Variation of Stress – Strain Properties with Rate of Strain and Temperature

In order to fully characterize the tensile properties of a material, the stress–strain relationships must be determined at both static and high-speed loading rates. Highly specialized apparatus and techniques are required to obtain the stress–strain properties over a wide range of strain rates from 20 to 2,000,000%/min; the time to rupture a specimen at these

TABLE 25

TENACITIES, ELONGATIONS, AND RUPTURE DENSITIES FOR VARIOUS FIBERS

Material	Rate of strain, %/min	Elongation at break, %	Tenacity at break, %	Rupture energy density, g-cm/den-m
High-tenacity nylon[a]	1	16.7	6.28	41.3
	10	17.6	6.76	50.1
	100	16.1	6.97	46.2
	300,000	14.7	7.57	43.8
	400,000	—	—	32.7
Fortisan[b]	1	5.4	6.33	17.6
	10	5.4	6.80	19.3
	100	5.3	7.04	20.4
	120,000	5.2	9.10	25.0
	350,000	—	—	18.0

Fiberglas[c]	1	2.8	4.73	7.6
	10	2.5	5.58	7.5
	100	2.4	6.07	7.7
	60,000	1.8	6.12	5.1
	200,000	—	—	5.1
Regular-tenacity nylon[d]	10	27.4	7.65	94
	60	27.1	6.66	93
	200	27.0	6.65	88
Polypropylene[e]	20	38.0	5.7	105
	240	32.2	6.0	98
	200,000	28.0	6.9	92
	370,000	27.0	8.0	94
	500,000	28.4	8.4	95

[a] High-tenacity nylon—Du Pont, type 300, 1085 den total.
[b] Fortisan—high-tenacity deacetylated cellulose acetate yarn, 787 den total.
[c] Fiberglas—type 150–1/0, 204 filament, 323 den.
[d] Regular-tenacity nylon—Nylon 66, 840 den, type 300 bright yarn.
[e] Isotactic polypropylene— 230–den, 70 filament yarn.

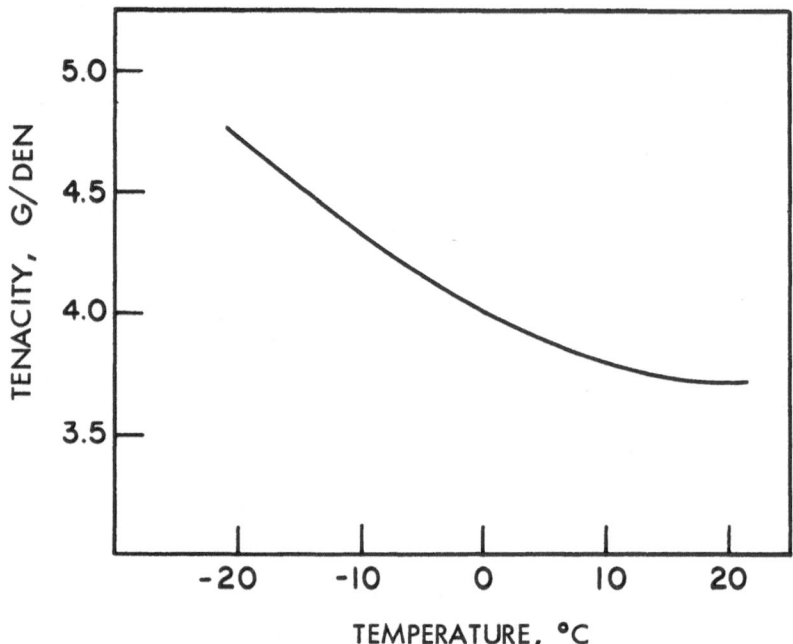

Figure 38.   Influence of Temperature on the Knot Strength of
Polypropylene Monofilament [31].

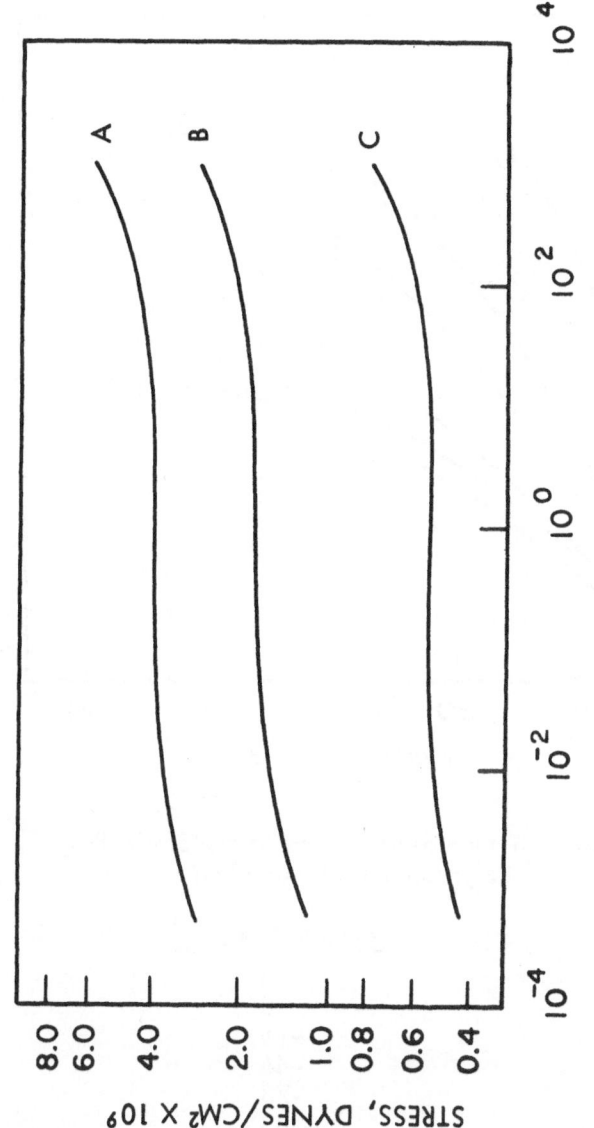

Figure 39. The Effect of Strain Rate on Stress at Given Elongations in 72-den Isotactic Polypropylene Filament [17]: (A) 14% Extension. (B) 5% Extension. (C) 1% Extension.

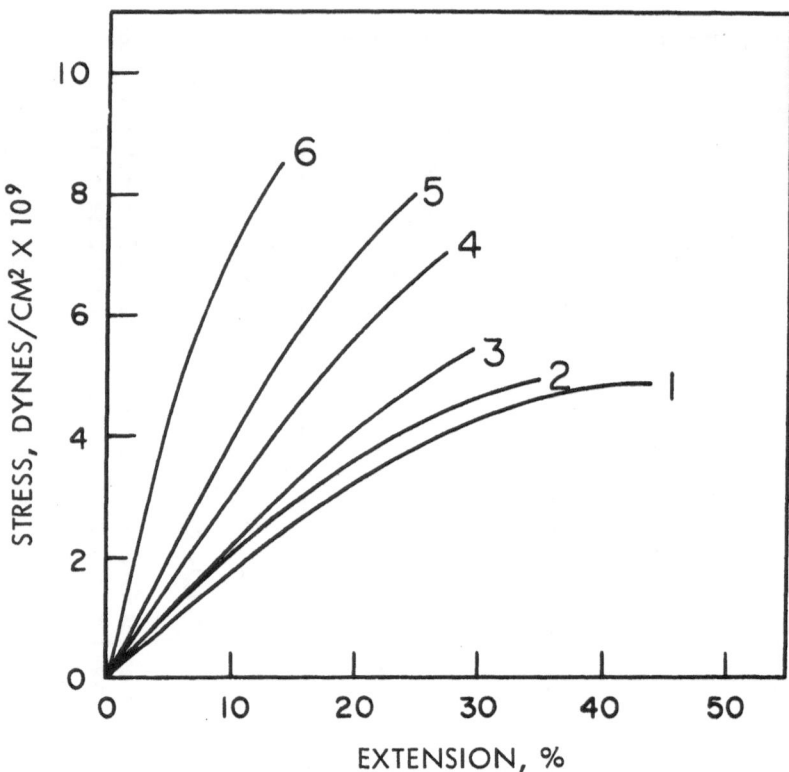

Figure 40. Effect of Temperature on the Stress-Strain Properties of Polypropylene Fiber [17]:

Curve	Temperature, °C
1	5
2	0
3	-16
4	-46
5	-73
6	-183

extreme strain rates is usually less than a thousandth of a second.

Table 25 presents comparative data on the tenacity, elongation, and energy absorption at various strain rates for nylon (high and regular tenacity), Fortisan (high-tenacity cellulose acetate), fiberglass, and polypropylene. The table shows that polypropylene exhibits the qualities for an outstanding, all-purpose fiber. Fiberglass, with an elastic modulus of about 300 g/den, is by far the strongest fiber considered; however, its breaking elongation of only 2–3% severely limits all applications for fiberglass except those requiring very small deformations. Over the wide range of strain rates tested, polypropylene exhibits outstanding energy absorption values, even exceeding those of regular-tenacity nylon; the breaking tenacity of polypropylene is slightly lower than Fortisan but comparable to that of nylon. The data in Table 25 show that the tenacity of fibers increases and rupture elongation decreases with increasing strain rates. These changes result from the fact that at the lower speeds of loading the viscoelasticity of the fiber permits partial deformation rather than immediate failure; at the ballistic speeds of loadings, the creep elongation is negligible, resulting in lower fiber extensibility.

Additional information on the stress–strain properties of polypropylene fiber under static and impact loading conditions is shown in Figures 39 and 40; the variation of strain rate on stress at given strains is plotted in Figure 39, while Figure 40 shows the effect of temperature on the stress–strain properties of polypropylene fiber. The curves in Figure 39 indicate that the stress is significantly dependent upon the strain rate at both extremes of the rupture time, but relatively unaffected by the intermediate strain rates corresponding from 10^{-1} to $10^{0.5}$ sec^{-1}. The predominant characteristic of the family of curves in Figure 40 is the rapid decrease in rupture elongation as the temperature is decreased from 5° to 16°C.

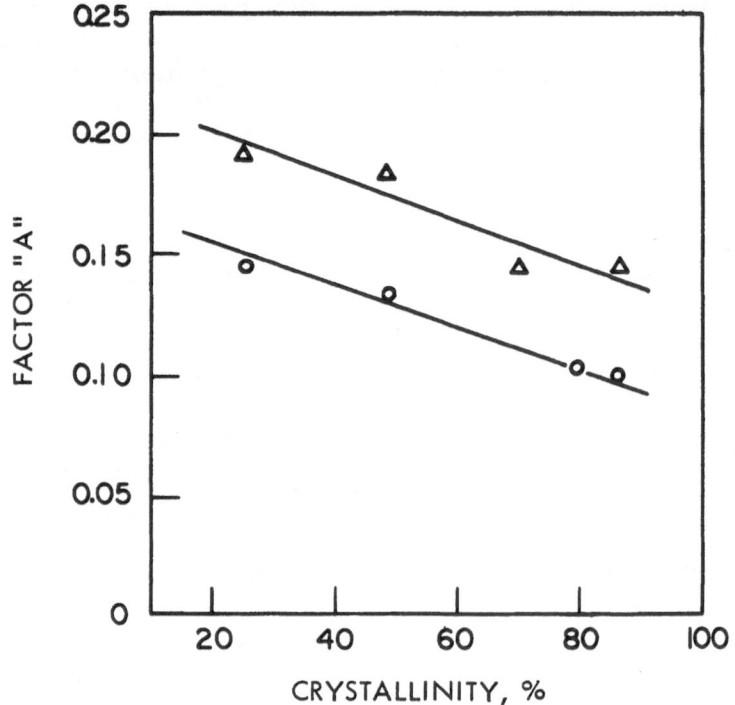

Figure 41. Dependence of the Factor "A" upon Crystallinity and Orientation[3]: (Δ) Slightly Oriented. (O) Highly Oriented.

5. STRESS – RECOVERY PROPERTIES

a. General Considerations

The above discussion on the stress–strain relationships of fibers is based on the application of a stress which ruptures the specimen, i.e., the fiber is loaded only once; in practice, however, fibers are seldom subjected to breaking loads, but rather to cyclic stresses which are small relative to the rupture load values. Consequently, important properties of fibers not evaluated by stress–strain diagrams are the elongation-recovery characteristics upon exposure to repeated stresses considerably smaller than the breaking load. These characteristics dictate the extent of wrinkle resistance, dimensional stability, and abrasion resistance exhibited by the fiber.

On removal of a stress which has produced an elongation below the rupture value of the specimen, a recovery toward the original fiber dimension will result, the extent depending upon the inherent properties of the material under test. The elongation resulting from the stress consists of (1) immediate elongation of the fiber and (2) additional creep elongation. The recovery (or lack of it) can be divided into the following components: (1) immediate elastic recovery (IER), (2) delayed recovery or primary creep, and (3) permanent set or secondary creep. The immediate elastic recovery is equal to the immediate elongation; however, either a portion or all of the stretch creep is recoverable; primary creep is the amount recovered, and the difference between the stretch creep and the primary creep is the permanent set. The immediate elongation is the elastic part of the stress–strain diagram in which the elongation is proportional to the load; the creep stretch is the nonlinear portion which can be considered as incomplete elasticity.

Cappuccio et al. developed an empirical relationship expressing the stress relaxation of polypropylene at 25°C, which is presented below:

$$\gamma/\gamma_1 = 1 - A \log t$$

where γ_1 is the stress after 1 sec, γ is the stress at time t,

TABLE 26

COMPARATIVE RECOVERY PROPERTIES OF VARIOUS FIBERS SUBJECTED TO 1 TO 2% STRAIN FOR ½ TO 6 MIN

Fiber	Recovery, %		
	Immediate	Delayed	Permanent set

A. Single-Fiber Recovery Properties After Holding at 2% Extension for 30 sec (Extension and Recovery at 10%/min)

Fiber	Immediate	Delayed	Permanent set
Polypropylene	91	9	0
Terylene, N type	95	5	0
Nylon 66	88	12	0
Tricel	84 to 95	5 to 16	-2 to 2
Orlon, type 81	60	24	16
Viscose	46	11	43

B. Single-Fiber Recovery Properties After Holding at 2% Extension for 3 min

Fiber	Immediate	Delayed	Permanent set
Polypropylene	82	18	0
Terylene, N type	91	9	0
Nylon 66	79	21	0
Tricel	63 to 73	11 to 19	about 16
Orlon, type 81	50	23	27
Viscose	34	29	37 variable

C. Single-Fiber Recovery Properties After Holding Under 1% Bending Strain for 6 min

	After 1 sec	After 100 sec
Terylene	83 to 86	93 to 96
Polypropylene	40 to 50	90 to 94

TABLE 27

RECOVERY PROPERTIES OF POLYPROPYLENE, NYLON, AND POLYACRYLIC FIBERS AT
STRAINS OF 5, 10, AND 15%

	Polypropylene	Nylon 66	Polyacrylic
5% Elongation			
Immediate elastic recovery	38.4	17.2	20.8
Delayed elastic recovery	61.6	82.8	73.7
Permanent set	0	0	5.5
10% Elongation			
Immediate elastic recovery	29.4	14.7	11.8
Delayed elastic recovery	64.2	79.9	56.4
Permanent set	6.4	5.4	31.8
15% Elongation			
Immediate elastic recovery	27.5	14.4	9.2
Delayed elastic recovery	61.7	71.0	49
Permanent set	10.8	14.6	41.8

TABLE 28

ELASTIC BEHAVIOR OF POLYPROPYLENE MONOFILAMENTS
UNDER CYCLIC APPLICATIONS OF STRESSES

Elastic parameter	Applied load, % of the rupture stress		
	25%	50%	90%
Breaking stress	100	100	96
Breaking elongation	100	93	92
Elastic modulus	93	66	52
Permanent deformation (% of the specimen length)	0	0.5	3.5

t is the time in seconds, and A is a factor dependent upon fiber crystallinity and orientation. The elastic recovery properties decrease as the value of A increases, i.e., the decrease of the stress with time is accelerated as A becomes larger. The dependence of the factor A upon crystallinity and orientation is illustrated in Figure 41; the plot shows that the elastic recovery increases considerably as the crystallinity and orientation levels increase.

b. Tensile Load – Recovery Characteristics

Quantitative data on the elongation–recovery properties of various fibers are contained in the following tables and figures: Table 26 presents comparative recovery properties of fibers subjected to 1 to 2% strain for 0.5 to 6 min. These data show that the recovery of polypropylene is complete for short stress times (0.5 to 3 min) and low strain rates of 1–2%; however, as the load time is increased to 6 min, the recovery of polypropylene is slower than that of Terylene polyester. Additional recovery properties of polypropylene, nylon, and polyacrylic fibers at strains of 5, 10, and 15% are given in Table 27; the load times for these data are unavailable. The permanent set exhibited by polypropylene and nylon 66 at the various strain rates is comparable; the permanent set in polyacrylic fibers is 4–5 times that of polypropylene at 10 and 15% elongation.

The effect of cyclic stresses on the mechanical properties of polypropylene monofilament is illustrated by the data in Table 28. The fiber was loaded for five continuous cycles with a stress between zero load and loads equal to 25, 50, and 90% of the rupture value; the elongation and recovery speed was 30 cm/min. These data show that the permanent deformation for the 25 and 50% loading is negligible, but increases to 3.5% (of the length of the test specimen) at the 90% stress. Figure 42 shows the effect of many (up to 500) cyclic stresses upon the permanent deformation of polypropylene fiber; the applied load was equal to 50% of the breaking stress. Under such a repeated stress, the permanent set increases rapidly to 1.5% after 100 cycles, then increases more slowly to 2.0% up to 500 stress cycles.

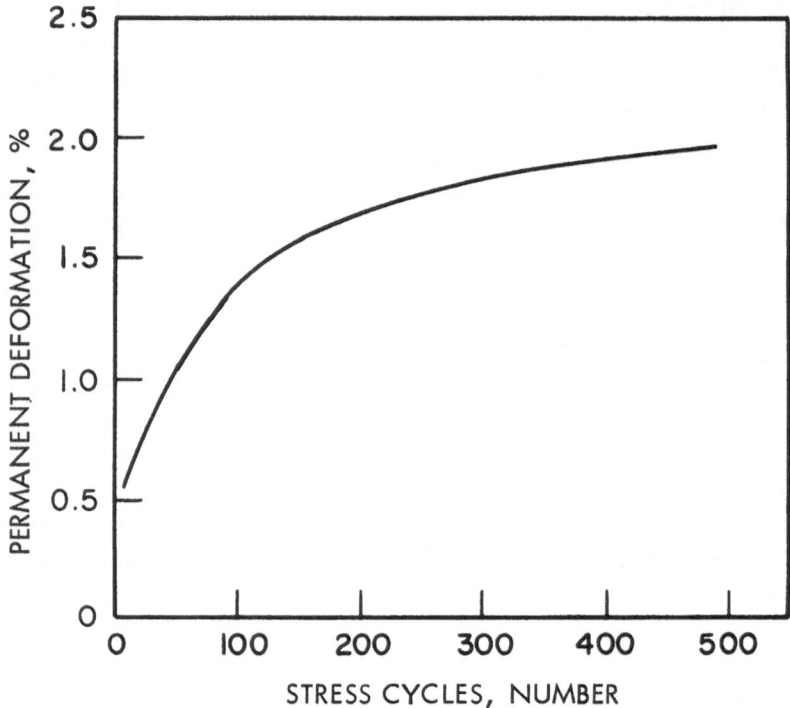

Figure 42. Variation of the Permanent Deformation of Polypropylene Fibers upon Cyclic Loading with a Stress Equal to
50% of the Rupture Stress [5].

Deformation and recovery properties of polypropylene fibers stressed for relatively long periods of time under small loads relative to the rupture values are shown in Figure 43. The fibers were stressed at 0.5, 1.0, and 1.5 g/den for 24 hr, after which the stresses were removed and the fibers then allowed a recovery period of 24 hr. Polypropylene recovers completely from loads of 0.5 and 1.0 g/den within 2 and 12 hr, respectively; the specimens stressed at 1.5 g/den for 24 hr exhibit a permanent deformation of about 1.5%. The instantaneous elastic recovery is the amount of stretch recovered upon immediate removal of the stress, i.e., at zero recovery time, as shown below:

Stress, g/den	Percent elongation after 24 hr of stress	Percent instantaneous recovery upon immediate removal of stress
1.5	7.0	4.75
1.0	5.0	3.25
0.5	2.5	1.00

Comparison of the growth characteristics between polypropylene and nylon 6 under a 1 g/den load for a period of 1000 min is presented in Figure 44, which shows that polypropylene fiber exhibits a resistance to elongation under load that is superior to nylon 6 yarn; the growth of polypropylene is slightly over 2% after 1000 min under a 1 g/den load, while nylon shows an extension of over 4%.

c. Compressional Load – Resilience Characteristics

A property similar to the recovery–load characteristics of a fiber is called resilience; while recovery is the ability of a material to return to its original dimension upon removal of a tensile load, resilience is a measure of fiber elasticity after exposure to a compressional load. The discussion above on tensile load–elongation characteristics with regard to instantaneous, creep, and permanent set phenomena also applies

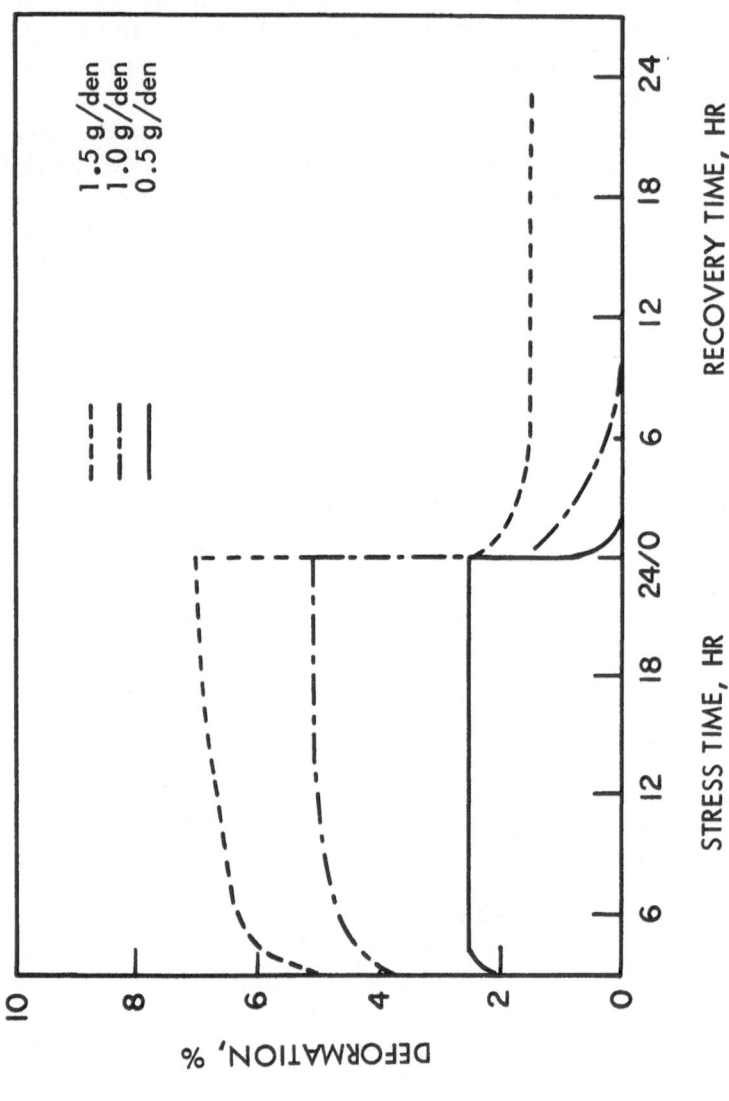

Figure 43. Deformation and Recovery Properties of Polypropylene Fibers Stressed at 0.5, 1.0, and 1.5 g/den for 24 hr [5].

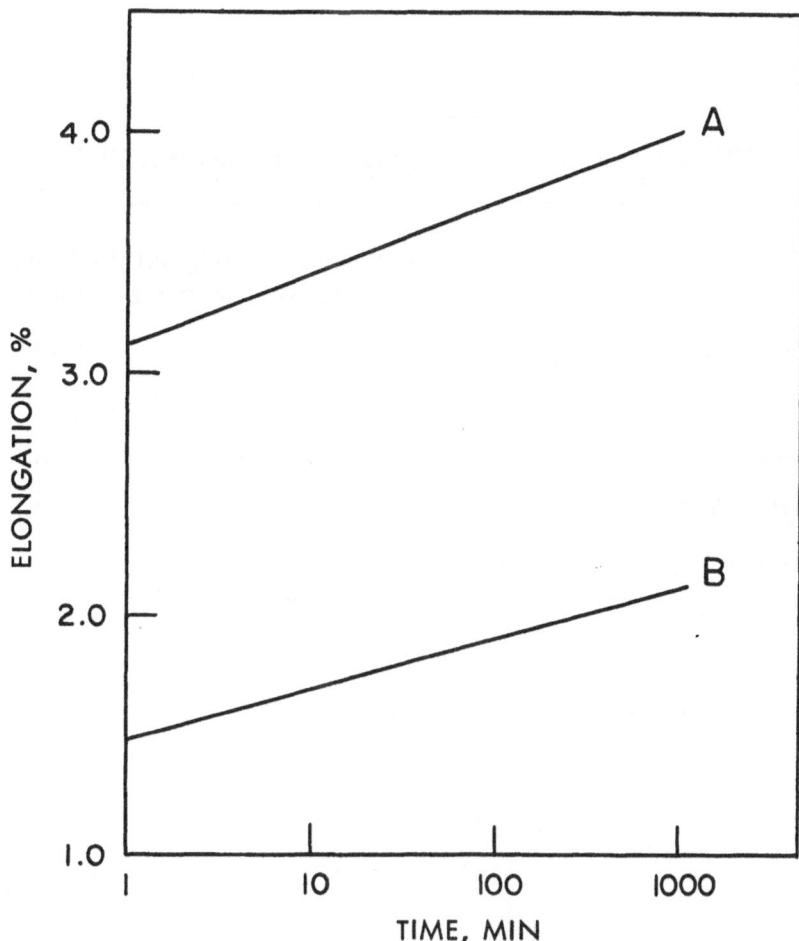

Figure 44. Elongation Characteristics of Nylon 6 and Polypropylene Fibers under a 1 g/den Load [21]: (A)Nylon 6.
(B) Polypropylene.

TABLE 29

TUFTED CARPET PILE RECOVERY PROPERTIES

Pile fibers	Percentage of original thickness after stated number of footsteps		
	2000	5000	10,000
Nylon	82	76	73
Wool	71	62	58
50/50 Wool/viscose	63	53	50
50/50 Wool /polypropylene	74	72	70

to fibers and fabrics in compression. The ability of a fiber to return to its original dimension after repeated compression is a function of the energy absorption characteristics of the material. A very common application in which resilience is of paramount importance is carpeting; consequently, much of the following discussion pertains to properties of fibers in the form of carpet pile.

The behavior of polypropylene fiber under compressional loads is illustrated by the following tables: Table 29 compares the recovery of various pile fabrics as a percentage of the original thickness after 2000, 5000, and 10,000 footsteps. The incorporation of 50% polypropylene with wool increases the recovery from 58% for a 100% wool pile to 70% recovery after 10,000 footsteps; the recovery of 100% nylon pile is 73%.

Table 30 compares the recovery of 100% wool, 100% polypropylene, and 50/50 polypropylene/wool carpets in terms of depression (inches) and percent pile height retained after 180,000 footsteps. These data further show the superiority of polypropylene to wool upon compressional loading.

Table 31 shows the comparative resilience between polypropylene and polyester fiber under cyclic compression and prolonged loadings. Under compression loads of 1 and 10 cycles, the recovery of polypropylene is comparable to the polyester; however, upon an extended load exposure of 1 hr, polypropylene shows slow recovery properties.

In summary, on the resistance to compressional loading, polypropylene exhibits excellent recovery properties that are very similar to nylon when the loading is cyclic and time is allowed for recovery such as encountered in carpeting. In tightly constructed fabric blends of polypropylene with other materials like wool, interfiber entanglement occurs which seriously hinders the recovery properties of the polypropylene. In addition to offering high strength and durability, polypropylene is considered exceptionally resistant to pilling—an undesirable accumulation of balls of fiber; pilling is prevalent in other strong fibers such as nylon and Terylene polyester.

TABLE 30

RECOVERY PROPERTIES OF 100% WOOL, 100% POLYPROPYLENE, AND 50/50 POLYPROPYLENE/WOOL CARPETS

Carpet	Depression, in.		Pile height, in. at $^3/_4$ lb/in.		Percent pile height retained
	Unworn	After 180,000 footsteps	Unworn	After 180,000 footsteps	
100% Polypropylene	0.152	0.053	0.364	0.274	75
50/50 Polypropylene/wool	0.162	0.064	0.384	0.277	72
Wool	0.132	0.057	0.327	0.245	75

TABLE 31

COMPARATIVE RESILIENCE OF POLYPROPYLENE AND POLYESTER FIBERS
UNDER CYCLIC COMPRESSIONS AND A 1-hr STRESS WITH 1000-g LOAD

Number of cycles or duration of stress	Polyester		Herculon Polypropylene	
	Compression, %	Recovery, %	Compression, %	Recovery, %
1 Cycle	30.7	63.1	28.4	63.2
10 Cycles	30.3	46.0	31.3	45.2
1-hr Load	29.4	57.2	29.8	32.6
5-min Recovery (from 1-hr load)	29.5	66.4	33.2	49.7

TABLE 32

COMPARATIVE ABRASION RESISTANCE OF VARIOUS TWINES OF COMPARABLE
DIAMETER — ABRASION AGAINST TUNGSTEN CARBIDE EDGE (100-g LOAD)—
RUBS TO BREAK

	Nylon	Hemp	Linen	Cotton	Polyethylene	Ulstron Polypropylene
Undyed						
Tested dry	26,000	970	520	920	1,000	10,000
Tested wet	4,500	270	350	250	600	5,300
Dyed						
Tested dry	5,000	400	540	860	1,020	12,000
Tested wet	1,650	380	480	200	750	5,500

6. ABRASION RESISTANCE

The abrasion resistance of a fiber is actually a measure of the ability of the fiber to repeatedly absorb the energy of deformation. The abrasion-resistant fibers are capable of elongation under stress, followed by complete recovery upon removal of the load. As discussed earlier, these deformation and recovery characteristics of a fiber are determined by the relation among instantaneous elastic deflection, primary creep, and secondary creep. Hamburger lists five properties as prime requisites for abrasion-resistant fibers: (1) low elastic modulus, (2) large immediate elastic deflection, (3) high ratio of primary to secondary creep, (4) high rate of primary creep, and (5) high magnitude of primary creep.

Comparative abrasion-resistance data of various twines presented in Table 32 were obtained by measurement of the abrading cycles to effect fiber breakage. Prior to the advent of the newest fibers, such as polypropylene, nylon was by far the most outstanding fiber for resistance to abrasion. However, Table 32 shows that Ulstron polypropylene filament exhibits a degree of abrasion resistance which exceeds that of nylon in three of the four fiber categories tested. Wet, dry, dyed, and undyed specimens were tested; the nylon yarn was superior only in the dry-undyed state, while the polypropylene exhibited the highest resistance under the other fiber test conditions of (1) wet-undyed, (2) wet-dyed, and (3) dry-dyed. This superiority in the dyed and wet states strengthens the competitive position of polypropylene for abrasion-resistance applications, which, in general, require a dyed fiber (clothes) and/or exposure to moisture, such as in fish nets and tow ropes.

A significant increase in abrasion-resistance resulting from blending polypropylene with wool is shown in Figure 45; the plot shows the variation of blend composition with the cycles of abrasion sustained to rupture the sample. No. 500A emery paper was applied at 5 psi, reciprocating at 125 times per minute, as the abrading agent; the abrasion was measured on a Stoll Universal Wear Tester.

Along with resilience, abrasion resistance is one of the two most important properties of a carpet. Information on the

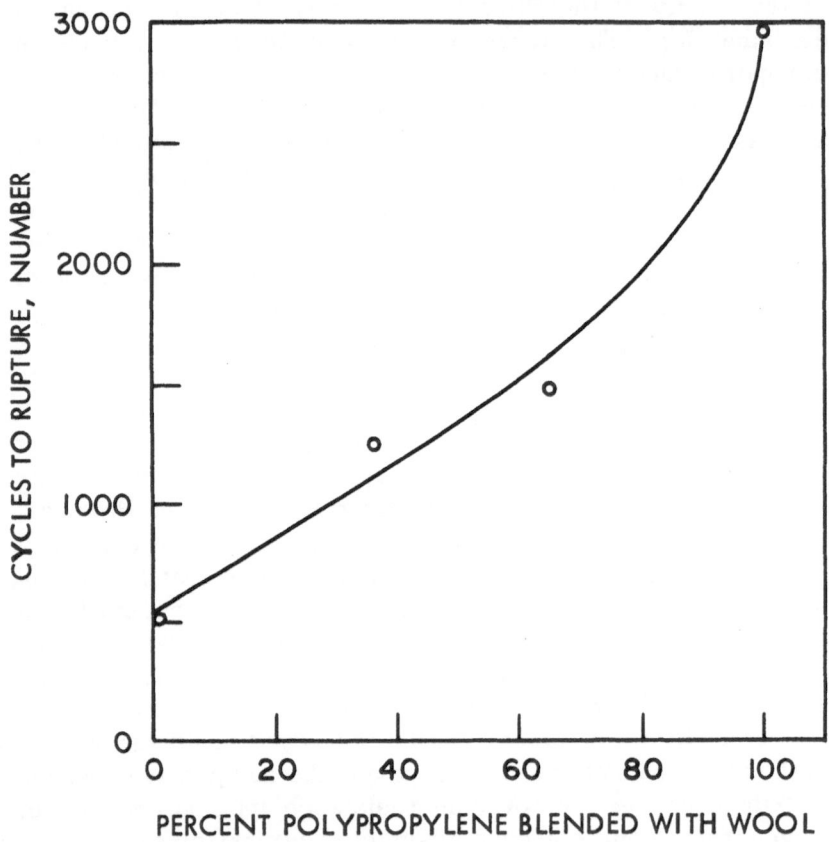

Figure 45. Effect of Polypropylene Fiber Content on the Abrasion Resistance of Wool—Polypropylene Blended Fabrics [38].

TABLE 33

THE ABRASION RESISTANCE OF CARPETS AS MEASURED BY THE TABER ABRASER TESTER

A. Taber Abraser Classification for Carpet Wear

Weight loss, mg, per 1000 cycles	Arbitrary classification
0–200	Excellent
200–450	Good
450–700	Fair
Above 700	Poor

B. Wear Resistance of Various Carpets

Fiber	Pile weight, oz/yd^2	Weight loss, mg, per 1000 cycles	Classification
Polypropylene	16–22	22–50	Excellent
Nylon	16–22	20–40	Excellent
Wool	16–22	400–450	Good
Viscose	28–32	1000–1500	Poor

TABLE 34

THE BURNING CHARACTERISTICS OF FIBERS

Fiber	Before touching flame	In flame	After leaving flame
Nylon 66	Melts before touching flame	Melts and burns	Does not readily support combustion
Nylon 6	Melts before touching flame	Melts and burns	Supports combustion with difficulty
Dacron	Melts before touching flame	Melts and burns	Burns readily
Dynel[a]	Shrinks away from flame and melts	Melts and burns slowly	Does not support combustion
Acrilan	Melts, ignites before reaching flame	Melts and burns rapidly	Burns readily with sputtering
Orlon 81, 42	Melts, ignites before reaching flame	Melts and burns	Burns readily with sputtering

Teflon	Melts when almost in flame	Melts and decomposes	Does not support combustion
Polyethylene	Melts, shrinks, and curls from flame	Melts and burns	Burns rapidly
Polystyrene	Melts, shrinks, and curls from flame	Melts and burns	Burns rapidly with production of a great deal of soot
Polypropylene	Shrinks rapidly from flame, curls, and melts	Melts, ignites with difficulty	Burns slowly
Asbestos	No effect	Glows	Does not burn
Glass	No effect	Glows	Does not burn

[a] Dynel— a copolymer of acrylonitrile and polyvinyl chloride.

TABLE 35

EFFECT OF POLYPROPYLENE BLENDS ON
THE FLAMMABILITY OF FABRICS

Fabric	Time of flame spread, sec
100% Rayon staple shirting 6.1	
90% Rayon–10% Herculon polypropylene 6.0	
80% Rayon–20% Herculon polypropylene 6.2	
70% Rayon–30% Herculon polypropylene 6.3	
65% Herculon–35% rayon 8.2	

abrasion resistance of polypropylene compared to other fibers has been developed during the evaluation of polypropylene in carpeting applications. Although the following information will be relative to the incorporation of polypropylene fibers in carpeting, the data are valid and useful for further illustration of the excellent abrasion-resistant properties of polypropylene.

Tread tests were conducted on 16-oz/yd^2 Axminister carpets consisting of (1) 100% wool, (2) 50% wool–50% polypropylene, and (3) 100% polypropylene. The carpets were laid on a stone staircase without an underfelt; the results of the tests were as follows: (1) The wool carpet was completely worn after 15,000 treads. (2) The 50–50 polypropylene–wool blend wore after 60,000 treads. (3) The polypropylene carpet showed no indication of wear after 70,000 treads.

The United States Rubber Company has used the Taber Abraser Tester to further evaluate the abrasion resistance of polypropylene fibers employed in carpeting. The Taber Tester subjects the sample to a rotary rubbing action under controlled conditions of pressure and abrasion. The degree of abrasion was measured by the sample weight loss per 1000 cycles with a CS-17 wheel in the Taber Tester. Test results on typical carpet fabrics are shown in Table 33 in terms of weight loss in milligrams per 1000 cycles for polypropylene, nylon, wool, and viscose fibers. These results show that polypropylene fibers have excellent abrasion resistance, comparable to that of nylon fibers. Also shown in Table 33 is an arbitrary classification for carpet wear based on the Taber Tester.

C. ENVIRONMENTAL BEHAVIOR

1. FLAMMABILITY

Comparative burning characteristics of various synthetic and mineral fibers are presented in Table 34. This information shows that synthetic fibers exhibit the following degrees of flammability: (1) burn rapidly (polyethylene), (2) burn slowly (polypropylene), (3) support combustion with difficulty (polyamides), and (4) do not support combustion.

TABLE 36

THERMAL SENSITIVITY OF VARIOUS FIBERS

Fiber	Temperature, °C		
	Soften	Melt	Decompose
Polypropylene	145–150	165	—
Polyethylene, low density	90–95	115	—
Polyethylene, high density	120–125	135	—
Dacron polyester	235	250	—
Orlon	235	250	—
Rayon	—	—	180
Acetate	200	230	—
Nylon 6	205	220	—
Wool	—	—	135
Silk	—	—	150
Cotton	—	—	150
Glass	735	—	—

From the data given above, polypropylene can be considered as intermediate in flame resistance as compared to other synthetic fibers. The beneficial effect of the addition of polypropylene to rayon (which burns rapidly) on the flammability of fabrics is apparent from the data in Table 35.

2. THERMAL STABILITY

This section is concerned with the effect of thermal exposure upon the dimensional stability of polypropylene fibers; the thermal effect on the mechanical properties of the fibers is found in the discussion of the specific property in an earlier section, "Viscoelastic Behavior."

As with most synthetics, polypropylene is a thermoplastic material which, as such, softens and becomes fluid (melts) under the influence of heat and pressure; in contrast, the natural fibers to not soften but decompose directly upon exposure to excessive heat. Because of this temperature sensitivity, it is possible to subject thermoplastic fibers to a thermal heat-set, a conditioning of the fibers which imparts a degree of dimensional stability against shrinkage that is dependent on the temperature of the heat-setting operation. For example, nylon fibers heat-set at 120°C (the fiber is maintained at a fixed length), then cooled and reheated to 120°C will exhibit less than 2% shrinkage; however, if the temperature is raised 180°C, about 5% shrinkage will result. If the fiber were not first heat-set, about 20% shrinkage would result. In addition to minimizing fiber shrinkage during use subsequent to manufacture, the heat-set treatment removes stresses generated during the various stages of fiber preparation; the removal of these stresses permits the production of flat, uniform fabrics without a tendency to curl or pucker.

a. Softening and Melting Temperatures

Quantitative data on the softening and melting points of various fibers are summarized in Table 36, which indicates that polypropylene is comparable to polyethylene and most natural fibers, such as silk, cotton, and wool, but significantly inferior to polyamides, polyesters, and rayon in regard to

TABLE 37a

THERMAL SHRINKAGE OF SYNTHETIC FIBERS AT 100°C

Yarn	Percent Shrinkage
Polyethylene, low density	40
Polyethylene, high density	10
Polypropylene	6

TABLE 37b

THERMAL SHRINKAGE OF POLYPROPYLENE FIBERS IN AIR AND WATER

Fiber	Percent Shrinkage		
	½ hr underwater 100°C (212°F)	½ hr in air at 100°C (212°F)	½ hr in air at 130°C (265°F)
Meraklon polypropylene	0.0	0.0	2.5
Nylon 66	0.0	0.0	1.4
Polyacrylic	0.0	0.0	0.5
Polyester	0.0	0.0	2.5

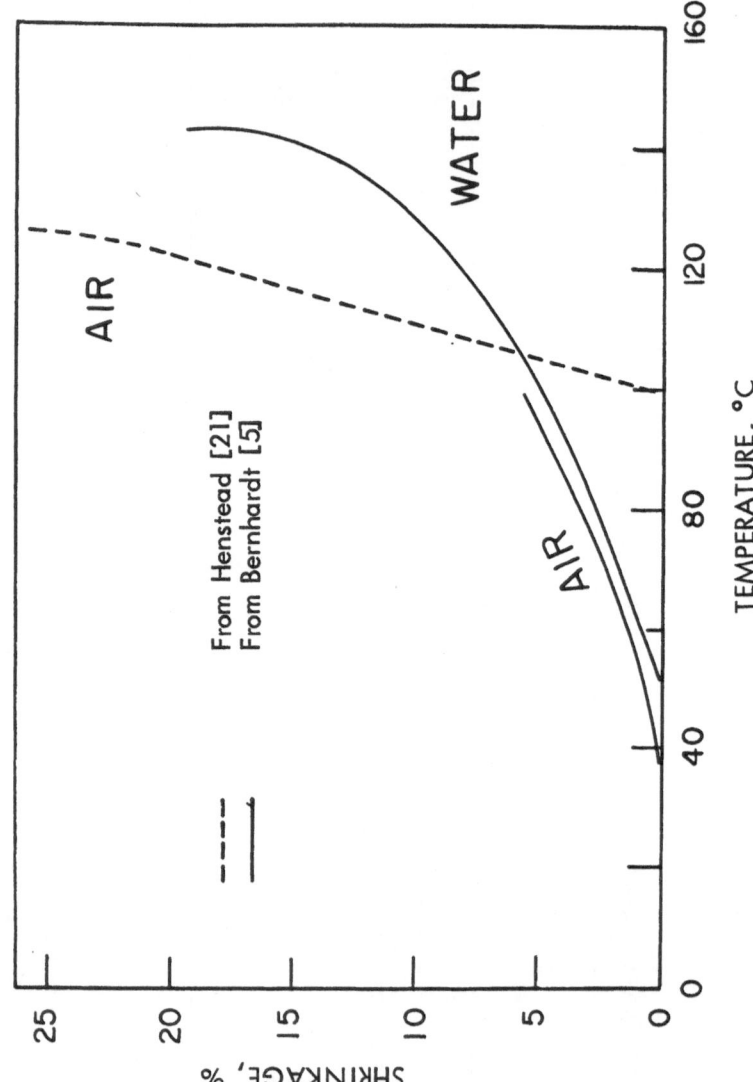

Figure 46.   Shrinkage of Polypropylene Fiber upon Thermal Exposure.

thermal resistance. The relatively low melting point of 165°C and sticking point of 145°C of polypropylene have presented some difficulties in providing a general-use fabric that can be pressed with a hot iron. However, many ironing tests conducted by the Hercules Powder Company indicate that polypropylene, in total or as a blend component in a fabric, may be ironed. Medium weight 100% Herculon fiber fabrics have been successfully ironed with a General Electric Company iron at the synthetic temperature setting of 120–138°C. Further tests showed that at the rayon temperature setting of 160–170°C, 100% Herculon fabric puckers after 1 sec under the iron; at this same temperature, a fabric blend of 65% Herculon–35% rayon resisted puckering for 10 sec of iron exposure.

b. Hot-Oven and Hot-Water Shrinkage

The percent shrinkage of polypropylene fiber as a function of fiber temperature and heating medium (air and water) is presented in Figure 46. The figure contains data from two investigators for fiber shrinkage under hot-air exposure; the data of Conti and Coen were obtained after a $\frac{1}{2}$-hr exposure of the fiber to the heating media. Table 37 compares the heat shrinkage of (1) yarns prepared from polyethylenes (low- and high-density forms) with polypropylene exposed to 100°C and (2) polypropylene (Meraklon) fibers with nylon, polyacrylic, and polyester fibers under hot-oven and hot-water exposures. The data in this table show that polypropylene exhibits excellent dimensional stability upon thermal exposure up to 212°F; in air at 265°F, the shrinkage of polypropylene is comparable to polyester but 2–5 times higher than nylon 66 and polyacrylic fibers. The significant effect of polypropylene in blended fabrics on the control of shrinkage upon washing is shown in Figures 47 and 48.

Figure 49 shows the stabilizing effect of heat-setting polypropylene fabrics with plots of percent shrinkage in treated (at 265°F) and nontreated samples upon thermal exposure. As mentioned above, and illustrated by the plot, heat-setting is a major factor in the control of fiber shrinkage. From the plot, shrinkage of the treated sample does not occur until the heat-

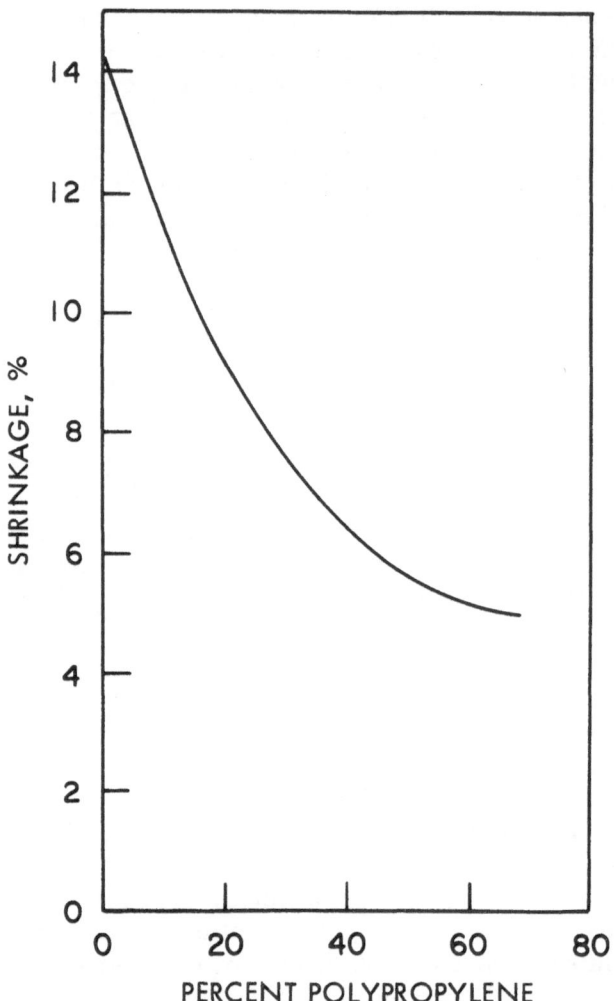

Figure 47. Effect of Polypropylene Fiber Content on the Laundering Shrinkage of Rayon – Polypropylene Blended Fabrics [21] .

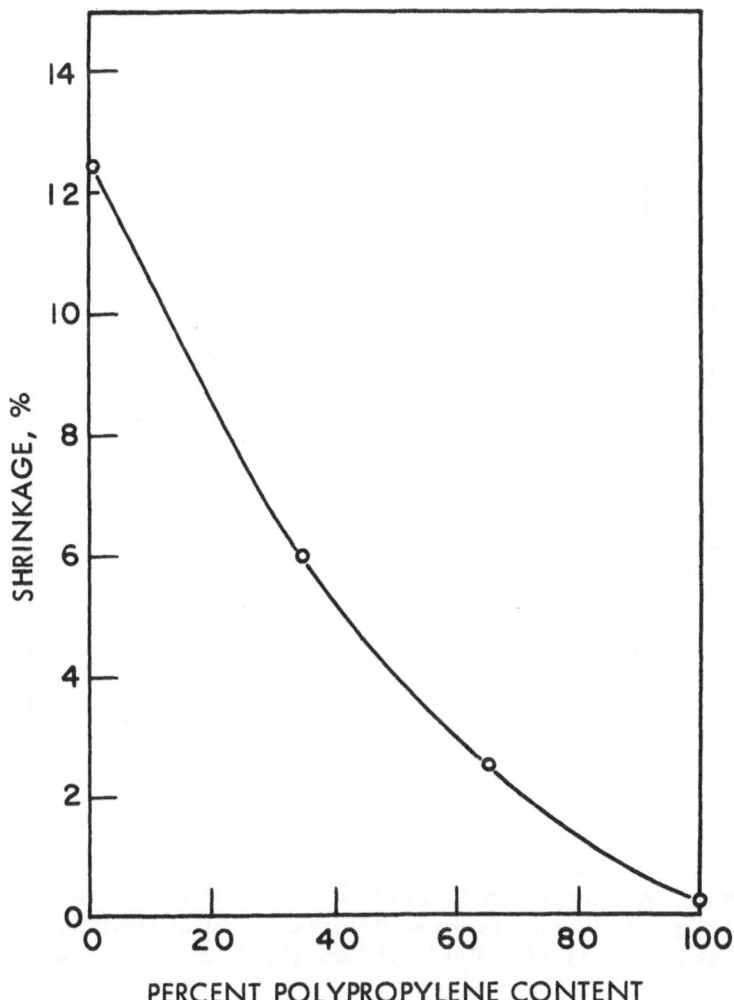

Figure 48.   Effect of Polypropylene Fiber Content on the Laundering Shrinkage of Wool – Polypropylene
Blended Fabrics [38].

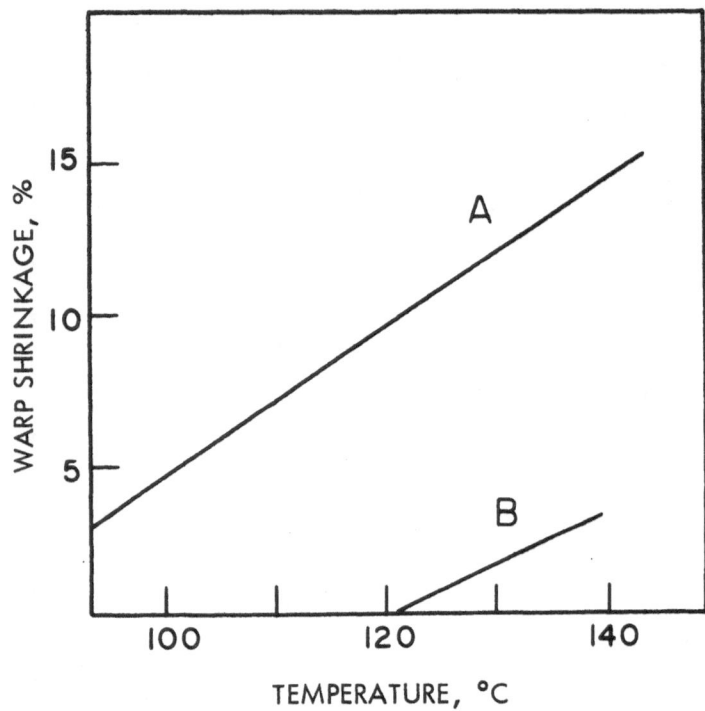

Figure 49.   Shrinkage Characteristics of 100% Polypropylene
Fabrics upon Thermal Exposure Before and   After Heat-Setting
Treatments [21]: (A)   No Heat Set. (B) Heat Set at 130°C.

set temperature is reached, while the untreated sample has already suffered about 10% shrinkage. Beyond the heat-set temperature of 260°F, the rate of percent shrinkage is comparable in both samples, but the absolute level is about five times lower in the treated sample.

3. DYEABILITY

One of the prime requisites which a fiber must fulfill, in order to be considered useful in the textile industry, is ease of dyeability. Since polypropylene possesses a compact molecular structure void of atomic groups with dye-attracting sites, the fiber lacks affinity for the absorption of conventional dyes, and consequently does not respond to the usual dyeing techniques. Since the discovery of polypropylene, manufacturers have extensively sought methods to increase the dye affinity of the fiber. The large number of recent patents and the great amount of literature on the dyeing of polypropylene serve as evidence of the worldwide research effort expended toward the development of new dyes and methods for dyeing synthetic fibers.

To date, several approaches have been developed which permit the preparation of a dyeable polypropylene fiber:

1. Copolymerization of the base monomer (propylene) with a small amount of a monomer containing active atomic groupings which readily accept usual dyestuffs.

2. The introduction of inert substances (added to the polypropylene prior to extrusion) which act as swelling agents, i.e., the resultant fibers become hydrophilic, which permits easy penetration of the dye liquor into the fiber. The influence of these agents on the permeability of dyestuffs into polypropylene fiber is probably the result of a distortion of the compact molecular structure of the polypropylene. This increased permeability is illustrated by the data of Dorset in Table 38, which shows that the moisture regain at 50, 80, and 100% relative humidity increases significantly for the modified fibers containing the foreign substances, compared to the untreated fibers.

Additional improvements in the above modified fibers are realized by further chemical and thermal treatments which not

TABLE 38

MOISTURE REGAIN FOR POLYPROPYLENE FIBERS
MODIFIED CHEMICALLY

% Relative humidity	Percent moisture absorption	
	Untreated fibers	Modified fibers
50	1.1	1.1
80	2.9	3.1
100	3.3	7.2

TABLE 39

MOISTURE REGAIN FOR POLYPROPYLENE FIBERS
MODIFIED CHEMICALLY AND THERMALLY

% Relative humidity	Percent moisture absorption	
	Untreated fibers	Modified fibers
50	1.1	1.8
80	2.9	3.6
100	3.3	43.5

only enhance fiber dye affinity but also provide substantial dimensional stability to the fibers. These treatments involve (a) dyeing the extruded fiber at elevated temperatures of 100°C or more and (b) use of additional chemicals such as monoethanolamine, formaldehyde, and polyethylenimine. The improved dye affinity of these further modified fibers is indicated qualitatively by the increased moisture absorption of the fibers as shown in the Table 39.

As mentioned in the section on light stability, polypropylene fibers are very adversely affected by ultraviolet irradiation. The additional benefit of stability toward ultraviolet light exposure imparted by these further treatments is shown graphically in Figure 50, which presents the difference in light stability of various treated polypropylene fibers as well as nylon fibers.

3. Probably the most promising approach to the dyeability problem is the development of special dyes which have an affinity for polypropylene fibers and can be introduced by conventional methods. Such a technique circumvents the more expensive and complicated methods described above. A recent patent describes certain dyes which can be conventionally applied and are effective in producing polypropylene fibers with good fastness properties. These special dyes, currently restricted to several colors only, contain long aliphatic chains which are effective in increasing the hydrophilic character of the fibers, resulting in adequate dye permeation. Examples of these dyes are benzene-substituted monoazo dyes which contain long-chained (seven carbon atoms minimum) aliphatic groups such as results from combining beta-naphthol with (1) diazotized para-amino-octylbenzene which gives a bright yellow dye and (2) diazotized octa-decyl-aniline which produces a bright yellowish-red dye.

The slight effect of dyeing on the properties of fibers is shown by the data in Table 40; data on the tensile properties for both dyed and undyed fibers, which include nylon and polyethylene in addition to polypropylene, are presented. A disperse dye of 3% Duranol Blue #2G300 was applied to the fibers for 1 hr at 100°C for these tests.

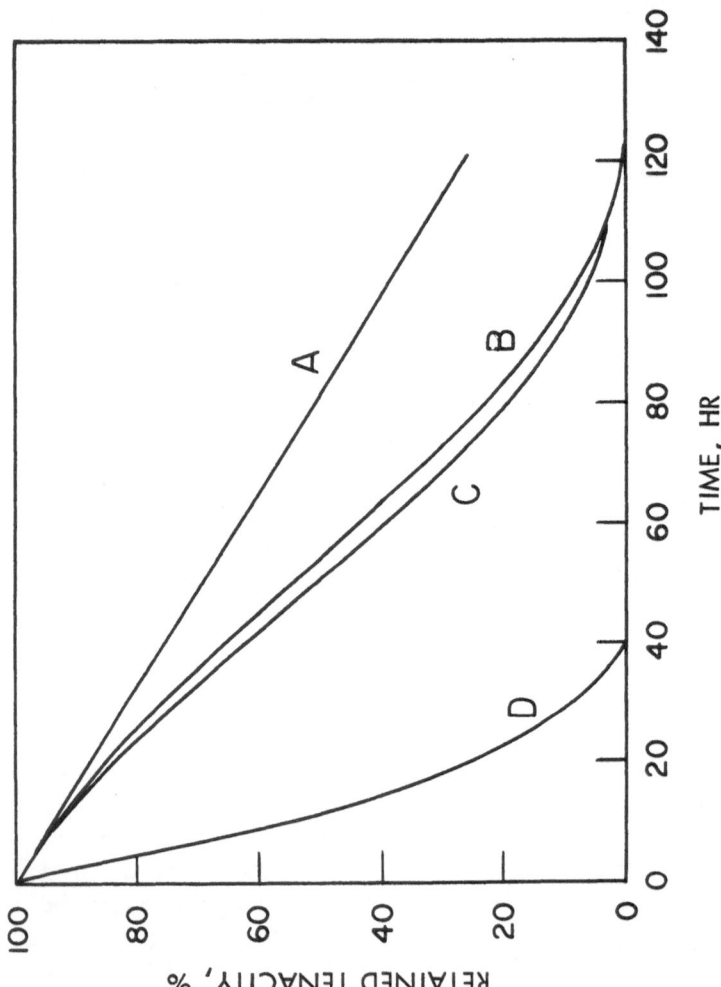

Figure 50. Resistance of Variously-Treated Fibers to Ultraviolet Light [9]: (A) Untreated Nylon 66 Fibers. (B) Fibers of Polypropylene and an Epoxy Resin Treated with Formaldehyde. (C) As (B) without Formaldehyde. (D) Untreated Polypropylene Fibers.

TABLE 40

EFFECT OF DYEING VARIOUS FIBERS WITH 3% DURANOL BLUE No. 2G300 FOR 1 hr AT 100°C

Parameter	Undyed			Dyed		
	Nylon	Terylene polyester	Ulstron polypropylene	Nylon	Terylene polyester	Ulstron polypropylene
Measured denier [a]	2002	2416	1901	2202	2563	2048
Breaking load, lb	27.0	26.6	30.4	25.8	26.4	31.3
Tenacity, g/den	6.1	5.0	7.3	5.3	4.7	7.0
Extension at break, %	18.8	13.4	22.4	30.9	22.2	29.1
Wet knot	14.4	16.4	18.8	14.1	13.8	18.3

[a] Length of 225 meters measured under atmospheric conditions.

TABLE 41

MOISTURE REGAIN OF VARIOUS FIBERS AT 70°F AND 65%
RELATIVE HUMIDITY

Fiber	Moisture regain, %
Cotton, raw	8.5
Wool	16.0
Acrilan acrylic	1.5
Orlon acrylic	1.5
Acetate	6.5
Teflon fluorocarbon	0
Glass	0
Nylon 6, regular	4–5
Nylon 66, regular	4.5
Nylon 6, high tenacity	4–5
Nylon 66, high tenacity	4.5
Polyethylene, high density	0
Polypropylene	0
Dacron polyester, regular	0.4–0.8

In spite of the above modifications and treatments, the commercial production of a dyeable polypropylene fiber has yet to be announced; cost and range of available effective pigments appear to be the major disadvantages of these modifications. Until the development of an economic and complete dye system, commercially colored polypropylene fibers will continue to be prepared via pigmentation prior to extrusion.

4. MOISTURE ABSORPTION

Strength, stiffness, and stability are usually considered as prime criteria for the evaluation of textile fibers. However, in addition to these qualities, fibers must also exhibit a certain degree of instability or workability—a factor which permits such textile operations as dyeing, heat-setting, and finishing. Next to temperature, moisture is recognized as the most influential agent on the workability of fibers.

a. Mechanism of Moisture Absorption in Fibers

The sensitivity of fibers to moisture effects (as well as thermal effects) is dependent on the composition as well as the geometric arrangement of the molecular structure of the fiber. As mentioned, fibers are composed of long, flexible macromolecules which, depending upon the structure regularity, are considered as crystalline, amorphous, or a combination thereof. The crystalline regions are characterized by well ordered and closely arranged molecular segments, while in amorphous areas the molecules are linked at infrequent intervals along the molecular chains, which results in an open and random structural arrangement.

The crystalline areas are practically immune to the penetration of moisture (as well as other liquids), while the amorphous regions, in sharp contrast, are readily attacked by permeants. Since the water molecules become attached to the fiber molecules, the degree of moisture absorption in fibers is also dependent upon the presence and availability of hydrophilic or polar groups in the amorphous regions. Some examples of these water binding groups in fibers are hydroxyl, amino, carboxyl, and carbonyl groups.

TABLE 42

SWELLING PROPERTIES OF FIBERS UPON IMMERSION IN WATER

Swelling properties	Length increase, %	Diameter increase, %	Cross-sectional area increase, %
Nylon 66, high tenacity	1.2	1.9–2.6	1.6–3.2
Polyethylene, low density	—	—	0
Dacron polyester, regular	0.1	0.3	0.6
Viscose rayon, regular	3–5	25–52	50–113
Polypropylene	—	—	0

b. Comparative Moisture Absorption of Fibers

A certain degree of moisture absorption is highly desirable in order to facilitate dyeing operations, as described previously. However, this same moisture is undesirable from the viewpoint of its deleterious effect on the physical dimensions, weight, and physical properties of textile fibers. The amount of moisture in fibers is usually expressed as percent moisture content or percent moisture regain. Percent moisture content is the weight of water calculated in terms of a percentage of the original sample weight, while percent moisture regain is based on the percentage of the dried fiber weight.

Table 41 contains moisture regains of various fibers under standard equilibrium conditions of temperature (70°F) and relative humidity (65%). The imperviousness of polypropylene to water, as seen from the table (zero moisture regain), results from the highly crystalline structure of the polymer and the absence of water-binding polar groups found in most other fibers. Because of this inertness to moisture, the physical properties of wet and dry polypropylene fiber are essentially identical. In contrast, the moisture-absorbing fibers can be easily dyed, but the moisture also produces harmful effects on the physical properties. An example of this undesirable aspect of moisture is shown in Table 42, which presents data on swelling properties of fibers upon immersion in water.

5. LIGHT STABILITY

As with other polyolefins, polypropylene is susceptible to the effects of ultraviolet irradiation, which adversely influences fiber strength and breaking elongation. The instability of polypropylene to radiation has been attributed to a decomposition of hydroperoxide groups formed at the tertiary carbon atoms via oxidation. However, by the incorporation of proper antioxidants and radiation absorbers, the light stability of polypropylene is greatly improved.

The effect of Florida and indoor laboratory aging on polypropylene fibers is illustrated by the following plots, in

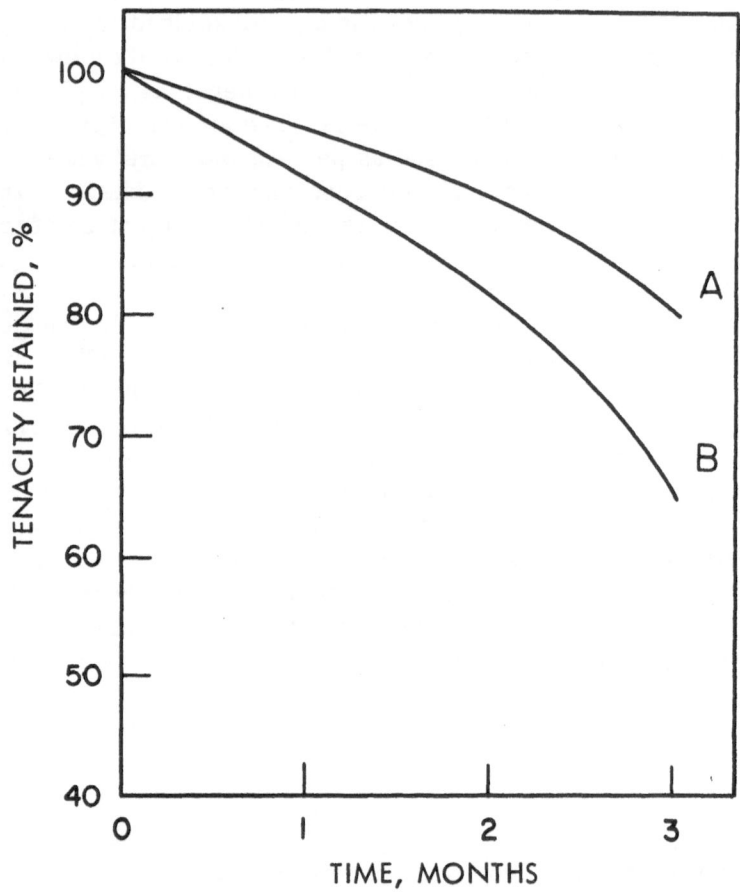

Figure 51. Resistance of Stabilized Polypropylene and Nylon 6
Fibers to Outdoor Exposure in Florida [21]: (A) Stabilized
Nylon 6 Fiber. (B) Stabilized Polypropylene Fiber.

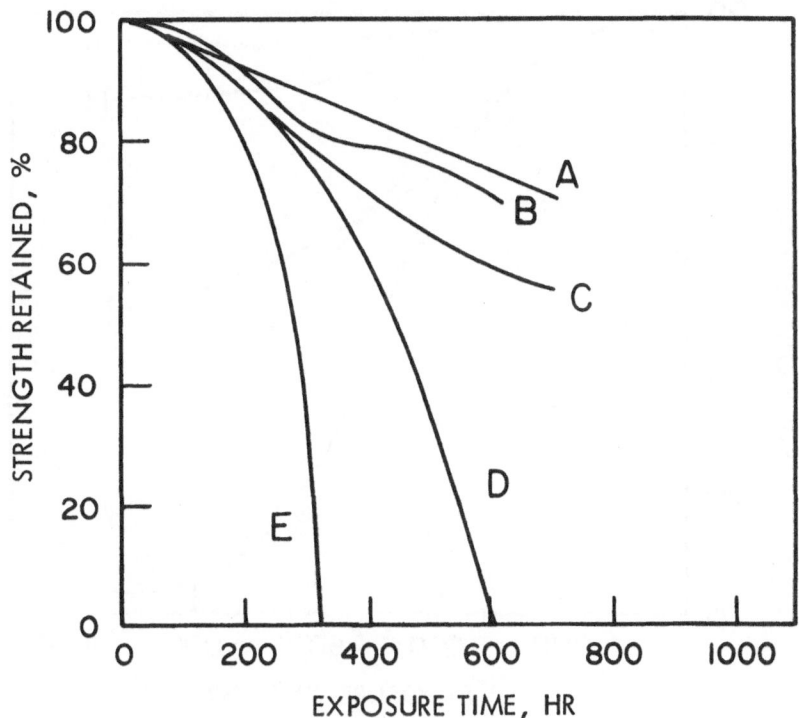

Figure 52. The Effect of Various Pigments on the Aging of Polypropylene Monofilaments upon Exposure to Florida Sunlight [30]:

Curve	Pigment
A	Black
B	White
C	Yellow
D	Red
E	Natural

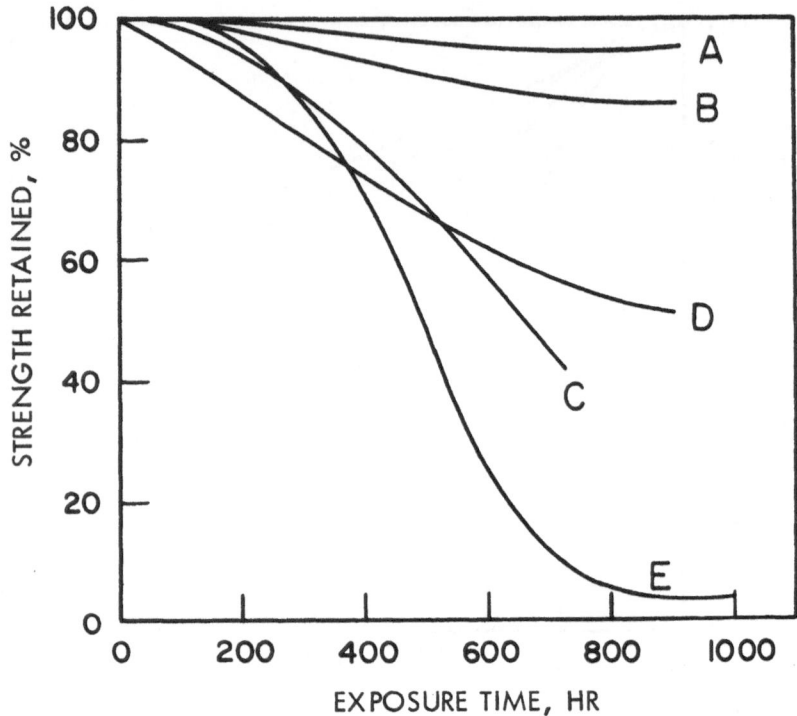

Figure 53. The Effect of Various Pigments on the Aging of Polypropylene Monofilaments upon Exposure to a 50°C Sunlamp [30]:

Curve	Pigment
A	Black
B	White
C	Yellow
D	Red
E	Natural

which the percent of original fiber strength retained is the measure of light stability: Figure 51 shows the comparative resistance of stabilized polypropylene and nylon 6 fibers upon exposure to Florida sunlight; although nylon exhibits superior light resistance, the stability of polypropylene is considered adequate for most outdoor applications.

Figures 52 and 53 are the result of a study by the Firestone Company to show the effect of various pigments on the aging of polypropylene monofilaments. This study indicates that black is the most effective stabilizer, followed by white, yellow, and red. The test specimens in Figures 52 and 53 were exposed to Florida sunlight and a sunlamp, respectively.

Figure 54 presents a family of curves showing the change in residual strength in terms of exposure time to a 100-W mercury vapor lamp for various denier polypropylene fibers as well as a dyed (black) fiber; the residual strength is expressed as a percentage of the initial strength. The lamp was placed in the focus of a parabolic mirror 11 cm from the monofilament. From the plots it is seen that the finer filaments are more sensitive to photochemical oxidation, while the addition of carbon pigment into the fiber imparts excellent resistance to ultraviolet irradiation. When a black color is undesirable, other, less effective, light stabilizers are available.

6. CHEMICAL RESISTANCE

The chemical resistance of polypropylene is exceptionally noteworthy. Extensive information on the chemical resistance of polypropylene fibers has been developed through exhaustive tests with literally hundreds of organic and inorganic chemicals. These data show that polypropylene fibers are very resistant to mineral acids, alkalies, aqueous solutions of inorganic salts, detergents, oils and greases, and organic solvents at room temperature.

In general, the chemicals that adversely affect polypropylene are organic solvents, such as chlorinated compounds, aromatic hydrocarbons, and the higher aliphatic hydrocarbons which, even at room temperature, produce swelling and softening; at

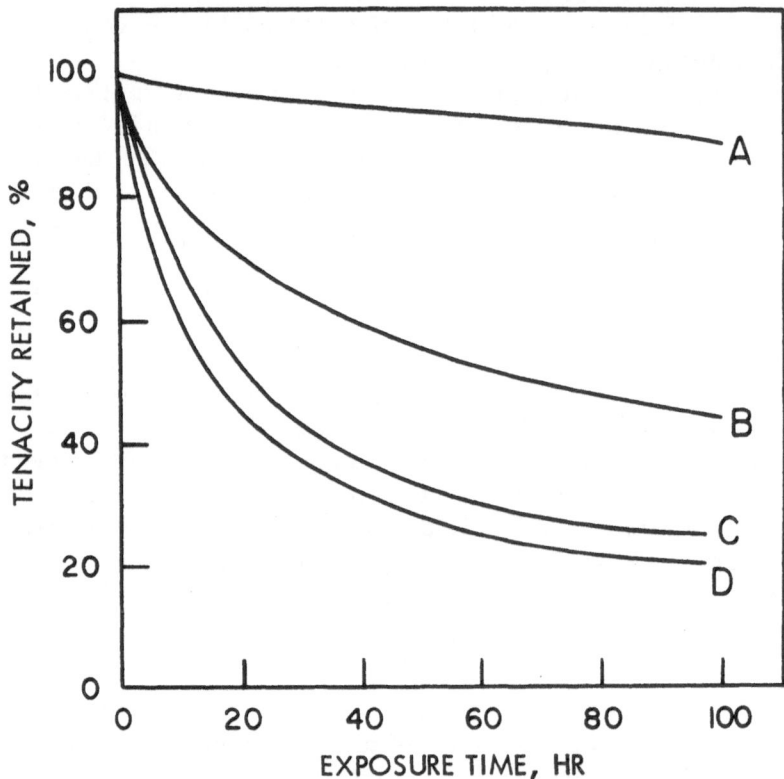

Figure 54. The Effect of Denier and Black Pigment on the Strength of Polypropylene Fibers upon Exposure to a 100-Watt Vapor Lamp [5]: (A) 260 den (Black Pigment). (B) 600 den. (C) 400 den. (D) 260 den.

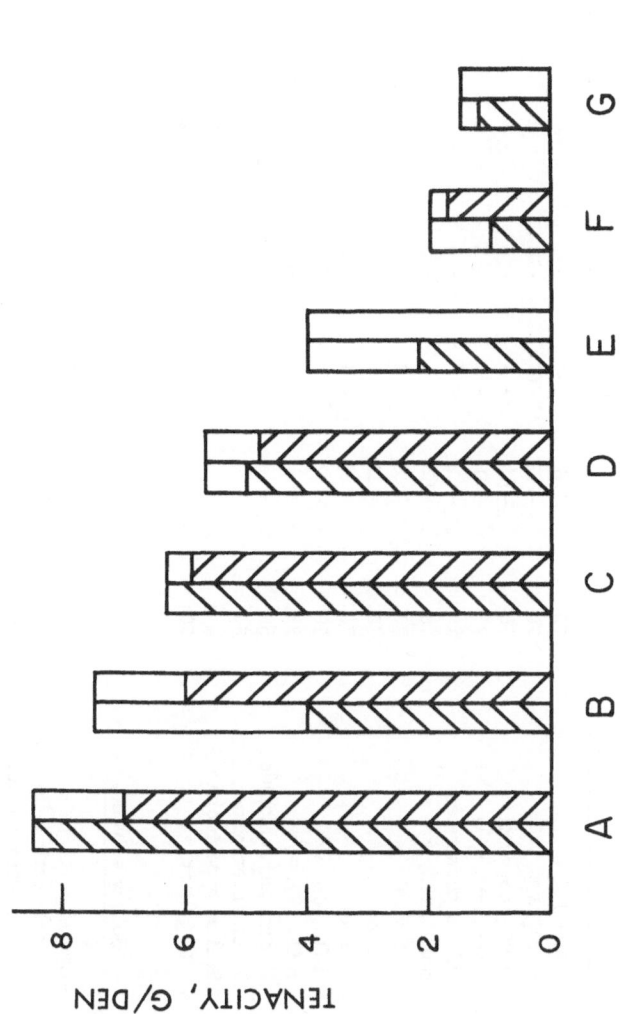

Figure 55. Chemical Resistance of Various Yarns to Solutions of 10% Sulfuric Acid and 40% Sodium Hydroxide at 70°C for 10 hr [38]: Yarns: (A) Polypropylene. (B) Nylon 66. (C) Terylene Polyester. (D) Orlon. (E) Viscose. (F) Cotton. (G) Cellulose Acetate.

TABLE 43

RESISTANCE OF 100% POLYPROPYLENE YARNS TO CHEMICALS

Reagent	Temperature, °C	Percent change after 7 days exposure				
		Weight	Denier	Rupture load	Tenacity	Extension
Control	70	Nil	—	—	—	—
H_2SO_4, 50%, w/w	70	-0.2	-13	-12	2	- 4
H_3PO_4, concentrated	70	-0.2	- 4	- 6	- 4	- 4
HNO_3, 20%, w/w	70	Nil	- 6	2	8	- 8
HCl, concentrated	70	-0.2	- 3	- 4	- 1	
NaOH, 20%, w/w	70	-0.3	1	- 3	- 4	- 8
m-Cresol	70	-0.8	2	2	1	- 5
Benzene	70	-1.2	7	- 1	- 8	23
Fuel oil	70	0.7	- 4	-13	- 9	- 8
H_2O_2, 10%, w/w	70	-1.2	Too weak to test			
H_2O_2, 10%, w/w	30	-0.1	- 6	-20	-17	- 2
Carbon tetrachloride	50	-1.0	10	2	- 7	21
Trichloroethylene	50	-1.1	- 3	- 4	- 1	- 5
Sodium hypochlorite, 10 g/liter	50	-1.1	1	13	12	2
Boiling solvents		Tested for 4 hr				
Benzene	boil	- 2.1	25	- 5	-24	75
Trichloroethylene	boil	-16.8	Heavy shrinkage, no tests			
Carbon tetrachloride	boil	- 3.9	52	Heavy shrinkage, no tests		
m-Cresol	100	- 0.8	46	26	-14	102

TABLE 44

COMPARATIVE CHEMICAL RESISTANCE OF POLYPROPYLENE AND POLYETHYLENE

Liquid	Polyethylene		Polypropylene, % crystallinity	
	Low density	Linear	63	56
A. 3 – 6 Weeks immersion				
Xylene	-30.7	-12.8	-5.3	-1.0
Turpentine	-37.7	-15.6	6.0	-16.3
Chloroform	-100	7.9	0.8	2.0
Hexane	9.3	23.5	-6.3	-14.8
Commercial bleach	-22.8	-18.2	-24.2	-20.9
1% Hydrogen peroxide	-13.9	-32.7	8.6	-7.2
B. 4 Months immersion				
Water	1.8	0	-1.1	-8.6
Isopropyl alcohol	1.1	-7.1	-7.1	-11.3
Primol D	-8.1	-78.0	10.1	-20.0
Silicone oil	-1.5	-11.0	13.8	-7.1
Methylethylketone	-5.2	-42.7	0	-6.0
10% Sodium hydroxide	-1.6	-14.0	3.2	-4.5
10% Common salt	2.6	-6.5	0	-9.1
10% Acetic acid	-8.1	9.8	13.2	-4.7
Dioctyl phthalate	-6.5	-2.3	5.9	-3.6
Linseed oil	-5.6	-60.5	1.1	-3.4
Corn oil	0	-50.5	16.0	-19.8
Methanol	2.2	2.5	2.3	-7.2
Igepal[a]	-100	-100	9.8	-1.8

[a] A surface-active agent used as a detergent.

elevated temperatures (160°F) polypropylene is soluble in these organics. The other substances which attack polypropylene are highly oxidizing reagents, such as fuming nitric acid, halogens, 100% oleum, and chlorosulfonic acid.

Quantitative data on the chemical resistance of polypropylene and other related materials are presented in the figure and tables discussed below: Figure 55 illustrates that the resistance of polypropylene fiber to both acids and alkalies is higher than various other synthetic fibers. Table 43 shows the chemical resistance of 100% polypropylene yarns in terms of percent changes in fiber properties. Table 44 compares the chemical resistance of polypropylene with polyethylene by noting the change in tensile strength upon exposure to the more active reagents. In the majority of the tests, the two types of polyethylene were more sensitive to exposure than the polypropylenes. In addition, since polypropylene has a higher initial tensile value, it may suffer a greater percent loss in strength and still be stronger than the polyethylene, showing a smaller percent loss upon chemical exposure.

V. Polypropylene Films

A. PROPERTIES OF PACKAGING FILMS

Polypropylene in its isotactic form became available in the United States in pilot-plant quantities in 1956 and was tested for molding and extruding, as a fiber and as a film. The basic properties of the polymer have been discussed in detail in the introduction, and in the first part of the book.

In the United States the largest application of film is in packaging, followed by electrical insulation and printed and decorative materials.

Table 45 gives properties of packaging films, and Table 46 similar data of pilot-plant material. Polypropylene film is known in two forms: unoriented and biaxially oriented. Biaxial orientation greatly improves the strength. Film made by chill-roll extrusion, which gives very little orientation across the sheet, has properties good enough to ensure a market.

Resins appearing exclusively as biaxially oriented film are those whose properties, when not so oriented, are inadequate to ensure markets. Polystyrene and polyesters are such materials. Other polymers, such as polyethylene, are improved by biaxial orientation, yet most polyethylene film methods do not take advantage of this. Polyethylene film, even when not fully oriented, has properties adequate for many uses. This is also true for cellulosics and vinyl films which are not biaxially oriented.

1. UNORIENTED FILMS

Two types of unoriented polypropylene film are made on chill-roll equipment. The polymer for Type 1 (see Table 47) has

TABLE 45

PROPERTIES OF PACKAGING FILMS

Base	Regenerated cellulose (cellophane)	Coated regenerated cellulose	Cellulose acetate
Forms	Sheets and rolls	Sheets and rolls	Sheets and rolls
Clarity	Transparent	Transparent	Transparent
Specific gravity	1.45	1.40–1.55	1.25–1.35
Thickness, mils	0.8–1.6	0.9–1.7	0.5–2
Maximum width, in.	50	60	52
Yield, in.^{2} of 1 mil film/lb	21,500	19,500	22,000
Tensile strength, psi	4000 – 18,000	4000 – 18,000	5000 – 12,000
Elongation, %	15–25	15–25	15–50
Elmendorf strength, g/mil	2–10	2–10	2–15
Folding endurance	Fair	Fair	Fair
Heat-sealing range, °F	Not sealable	200–300	350–450
Flammability	Slow burning	Slow burning	Slow burning
Water absorption in 24 hr immersion test, %	45–115	–	8–10
Dimensional change at which relative humidity, %	3–5	3–5	0.6–80
Water-vapor permeability, g/24 hr/100 in.^{2} at 100°F, 90% relative humidity	High	0.2–1.0	100
Permeability to gases, oxygen and carbon dioxide	Dry - low Moist - variable and higher		Medium
Resistance To alkalies	Poor to strong alkalies	Poor to strong alkalies	Poor to strong alkalies
To acids	Poor to strong acids	Poor to strong acids	Poor to strong acids
To greases and oils	Impermeable	Impermeable	Good
To solvents	Insoluble	Insoluble	Soluble, except in hydrocarbons
To sunlight	Good	Good	Good
Solubility, temperature, max °F	300	300	200
min° F	Depends on type and relative humidity		Becomes brittle

Rubber hydrochloride	Polyethylene	Vinylidene copolymers	Vinyl resins	PVC and nitrile rubber blends
Continuous rolls and sheets	Rolls and sheets, flat tubing, gusseted tubing	Seamless tubes, rolls	Rolls	Rolls, sheets, and tubing
Transparent	Translucent	Transparent	Transparent to hazy	
1.12–1.15	0.92	1.68	1.23–1.27	1.18
0.4–2.5	0.5–10	0.5–10	1–10	1–3
60	196	54	84	40
24,000	30,000	16,300	21,600	23,500
3500 – 5500	1500 – 2500	1800 – 15,000	1400 – 5500	2500 – 4000
350–500	50–600	20–140	150–500	250–500
20–1000	75–200	40	60–1000	300
High	High	High	Good	High
250–350	230–300	–	200–350	325–400
Nonflammable	Slow burning	Self extinguishing	Slow burning	Slow burning
5	0.005	Negligible	Negligible	Small
Slight	None	None	None	None
0.5–1.5	1.2	0.15	4–6.0	9.4
Low to high	High	Very low	Medium	Low
Good	Excellent	Good except ammonium hydroxide	Good	Good
Good	Excellent	Excellent except sulfuric and nitric acids	Good	Good
Good	May swell	Excellent	Good	Excellent
Soluble in cyclic hydrocarbons and chlorinated solvents	Good but may swell	Excellent	Soluble in some	Soluble in some
Fair	Good	Good	Fair	Fair
200	180	200	200	200
–20	–60	–20	–50	32

TABLE 46

PROPERTIES OF EXPERIMENTAL CLEAR POLYPROPYLENE FILMS

Property	Profax unoriented	Profax oriented	Polymer A unoriented	Polymer A oriented	Polymer B unoriented	Polymer B oriented
Thickness, mils	4.0	1.5	5.0	1.4	4.0	1.2
Yield, in.^{2}/lb/mil	31,000	31,000	31,000	31,000	31,000	31,000
Specific gravity	0.885	0.895	0.90	0.90	0.90	0.90
Tensile strength, psi	2500–4500	up to 20,000	4000	24,000	3500	30,000
Elongation, %	600	11,200±50	700	11,250±80	500	11,200±70
Folding endurance	High	High	High	High	High	High
Heat-sealing range, °F	350	~400	~350	~400	~350	~400
Water permeability	0.5–0.7	0.2–0.3	0.5–0.6	0.20–0.25	~0.5	~0.25
Gas permeability, cm^3/100 in.^{2} per mil thickness/24 hr at 100°F						
Oxygen	100	55	120	60	120	50
Carbon dioxide	250	200	280	220	260	200

a melt index of 0.2 and a high isotacticity, whereas that for Type 2 has a melt index of 4.0 and a lower degree of tacticity. Properties are shown in Table 47 for polypropylene, polyethylene, and cellophane.

The lower-melt-index film retains its strength at much lower temperatures than does the higher-melt-index film. Poor low-temperature properties are often cited as a weakness of polypropylene, and the higher-melt-index film should not be recommended for applications such as frozen-food wrapping, where low-temperature resistance is needed. The lower-melt-index film, however, is adequate for such purposes.

The stiffness or the tensile modulus is greater in the Type 1 material, while Type 2 polypropylene film is in the same range as medium-density polyethylene. Type 1 polypropylene approaches the values of high-density polyethylene or cellophane.

The range of physical properties of the film is set by conditions of extrusion. The stiff cellophane-like film or the flexible polyethylene-like film can be made from the same polymer under different extrusion conditions.

Stiff film will have poor tear and low impact strength, whereas the flexible film will be good in these properties. Property compromises, normally requiring many different polyethylenes, are possible with a few polypropylene species.

The number of square inches of 1-mil film of a pound of polymer is the area factor. Polypropylene has the highest area factor of any film available. Polypropylene will be cheaper for the same area than any other film at the same price per pound. This advantage is small over polyethylene but it is significant over cellophane.

The tensile strength of the selected samples is low compared to some reported for polypropylene film. The Type 1 material exceeds polyethylene and approaches cellophane. The high elongation at this tensile strength indicates that the polypropylene film is tougher than competitive films, even though stiffer than most of them.

Polypropylene is heat-sealable by the same equipment used for polyethylene, requiring about 50°F higher sealing temperature.

TABLE 47

PROPERTIES OF UNORIENTED POLYPROPYLENE FILMS

Property	Polypropylene		Polyethylene		Cellophane
	Type 1	Type 2	Medium density	High density	
Yield, in.²/lb/mil	29–31,000	29–31,000	29–30,000	29–30,000	18–20,000
Tensile strength, psi	2800–4400	2400–2600	2800–3100	3400–3600	4400–18,000
Elongation, %	200–250	300–375	75–225	10–300	15–45
Tensile modulus, psi · 10³	100–115	45–50	50–70	100–125	150–200
Spencer impact at 75°F, psi	350–450	1500–2000	1500–3500	1000–2000	4400–8100
Haze, %	2–3	2–3	5–10	5–10	1–3
Water-vapor permeability, g/100 in.² per 24 hr/mil at 100°F	0.65	0.65	0.7	0.25	3.5–6.8
Gas permeability, cc/100 in.² per atm/24 hr/mil at 75°F					
Oxygen	97	97	280	125	0.4
Carbon dioxide	280	280	990	580	0.05

Water and gas permeability of polypropylene film are similar to polyethylene. Cellophane is a better gas barrier, but in the uncoated form it has a higher moisture-vapor transmission.

2. ORIENTED FILMS

Table 48 presents data on the properties of an experimental biaxially oriented polypropylene film. The improvement in tensile strength over the unoriented film is evident by comparison with Table 47. The increase in strength is accomplished without any drastic reduction in elongation. Orientation in both directions is not the same, and properties differ in the two directions. The techniques for making such film are highly developed.

Film grade polypropylenes range in density from 0.880 to 0.900 and in melt index from 0.2 to 2.0. They must be free of remnants of catalyst and microgels. Their tacticity need not be as high as for the fiber grade, as highly isotactic polymers tend to give films which are not brilliantly transparent but either show some haze after casting or develop it on storage. The melting point of the best isotactic fractions is 176°C, with a density of 0.915 and a second-order transition temperature of -18°C; the thermogram of these materials shows no endotherm until 160°C and then a very sharp melting-point minimum. Film materials start with a drop of the thermograph curve around 150°C and give a somewhat broader endotherm with its minimum in the neighborhood of 168°C and a density around 0.99.

When polypropylene is substituted for polyethylene on equipment for blown tubing, film may readily be made, but it does not have gloss or clarity equal to polyethylene-blown tubing, except at thicknesses below 0.3 mil (7.5 μ). The physical properties are about equal to polyethylene.

B. MANUFACTURE AND PROCESSING CONDITIONS

Slots and dies of conventional type are usable for making polypropylene film by the chill-cast method with modifications

TABLE 48

PROPERTIES OF AN ORIENTED POLYPROPYLENE
(0.5 MIL)

	Along sheet	Across sheet
Tensile strength, psi		
Room temperature	15,000	60,000
300°F	4,000	8,800
Elongation at break at room temperature, %	200	35
Elmendorf tear strength, g	12	13
Mullen burst strength, psi	37	
Moisture-vapor transmission, g/100 in.^{2}/24 hr at 25°C	0.28	
Gas permeability, cc/100 in.^{2}/24 hr		
Oxygen at 25°C, 1 atm	200	
Carbon dioxide at 25°C, 1 atm	730	
Yield, in.^{2}/lb/mil	61,000	

	Along sheet	Across sheet
Shrinkage, %		
200°F	None	None
250°F	5	8
300°F	12	28

Specific gravity	0.885–0.895
Yield, in.^{2}/lb/mil (theoretical)	39,900–31,300
Water-vapor transmission rate, g/100 in.^{2}/24 hr/mil	0.45–1.10
Maximum use temperature before softening, °F	300
Types of heat seal	Impulse and radiant bar
Heat-sealing range, °F	350–425
Tensile strength (average), psi	
Machine direction	7600
Transverse direction	4600
Elongation (average), %	
Machine direction	741
Transverse direction	730
Elmendorf tear strength (average modulus), g	
Machine direction	82
Transverse direction	426
Gas permeability, cc/100 in.^{2}/mil/atm	
Oxygen	180–260
Carbon dioxide	586–870
Clarity, haze, gloss	Excellent optical properties
Dimensional stability	Excellent up to maximum use temperature

(cont.)

(TABLE 48, cont.)

Printability	Equivalent to polyethylene
Resistance	
To acids	Excellent
To alkalies	Excellent
To organics	Variable

	30°C	150°C
Power factor, 10^8 cycles	0.0011	0.0063
Dielectric constant, 60 cycles	2.11	1.86
Volume resistivity, Ω -cm	$4.9 \cdot 10^{14}$	$3.6 \cdot 10^{12}$
Surface resistivity at 25°C	$1.4 \cdot 10^{15}$	
Dielectric strength at 25°C	800 V/mil	

so that (1) die lips should be tapered for close approach to roll surface, (2) the land should be long with a gradual approach, (3) a solid jaw on the roll side prevents leaks, and (4) die passages must be streamlined. Extrusion dies must be stress-relieved of strains above the operating temperature of about 280°C. All contact surfaces must be finished and chrome-plated.

A low-velocity air stream forces the film against the chill roll shortly after it leaves the die. This air stream must be controlled to avoid flutter in the web.

In the "air-knife" technique an air duct is fitted with adjustable lips of the same width as the chill roll and with an opening of 0.050 in. Air is fed through the duct at 5 to 10 psi. The "knife" of air presses the molten polymer against the chill roll. The knife should be directed at the roll surface and should contact the web immediately after the film leaves the die.

Good contact with the chill roll prevents puckering. Longitudinal strips in the film are flat, and between these the film will be puckered or wavy. The effect is caused by poor contact with the chill roll, with unevenness in cooling rates over the film area.

The speeds at which these defects appear are higher for polypropylene than for polyethylene. With polyethylene, deposits build up rapidly at the die lips and are apparent even after only a few hours of operation. Polypropylene film dies may be operated for months without shutdown by die deposits. If die passages are not perfectly streamlined, or if the inner die surface is roughened in any manner, polyethylene builds up oxidized material within the die or in polymer passages elsewhere. This buildup periodically peels off, causing streaks in the film.

Polypropylene extrudes better at the highest screw speed and output rates. Buildup of pressure on the screw by various techniques does not improve the appearance, nor does it stabilize extruder operation.

A long extruder barrel is a greater advantage in polypropylene extrusion than in other thermoplastics. An extruder

TABLE 49

PROCESSING CONDITIONS

Parameter	Low melt index	High melt index
Temperature, °F		
Cylinder	450–550	450–500
Die	525–550	500–525
Stock	550–575	525–550
Chill roll	40–100	40–100
Screen pack	About 100 mesh	About 100 mesh
Extruder, length/diameter ratio	20/1	20/1
Die gap, mils	15–25	15–25
Die land, in.	0.8–2	0.5–1.5
Air gap, in.	$^1/_4$–$^1/_2$	$^1/_4$–$^1/_2$

with a 15:1 ratio of screw length to diameter will extrude less polypropylene than, say polyethylene, at any given speed. If the ratio of length to diameter is above 20:1, however, often the output of polypropylene will be greater than that of polyethylene. Polypropylene shows a sharper drop in viscosity as it heats up than do most other thermoplastics.

Extrusion temperature is from 525° to 575°F; below this the film is not clear and above 600°F the polymer degrades. Table 49 shows typical conditions. Surge is encountered in film extrusion. This appears as variations in thickness and as variation in width. The normal reaction is to slow down or employ a heavier screen pack. This makes the situation worse, since the problem does not originate in the screw but in the die. The low melt viscosity does not allow enough buildup of pressure in the die to establish a positive flow pattern.

Corrective action is to increase extruder speed, lengthen the die land, and decrease the die opening. Die lands of 1 in. or more are recommended, and lengths below 0.25 in. are almost inoperable. A die opening from 15 to 25 mils is satisfactory with a reasonably long land. Wider die openings orient the film in the machine direction and increase tearing along the direction of extrusion.

Biaxial orientation of polypropylene film draws the film out sidewise as well as in the direction of extrusion. Table 50 shows that stretching increases the strength of the film. The strength of polypropylene film made by the chill-roll method is adequate for many uses. Chill-roll film may receive a small amount of stretch across the web. While the film as a whole makes no firm contact with the chill roll, the edges do. They stick to the roll when they touch it if the chill-roll temperature is above 70°C. As the film cools, the edges remain in place, and the film between is oriented. The high linear expansion coefficient makes this stretch appreciable. Increasing web tension extends orientation in the longitudinal direction.

A textile tenter frame for control of the width of cloth consists of a series of clips arranged on two endless chains

TABLE 50

EFFECT OF ORIENTATION ON STRENGTH

Stretch, %	Ultimate tensile strength, psi	Elongation, %
None	5,600	500
200	8,400	250
400	14,000	115
600	22,400	40
900	23,800	40

which run on two horizontal tracks. The tracks can be arranged so that the clips grasp the edge of the film as it emerges. The tracks diverge slightly so that the clips pull the film sideways. The angle of the tracks can be regulated to any desired stretch. The tenter frame may be mounted in an oven to control the temperature at which orientation occurs. This gives biaxial orientation to Mylar and polystyrene and is applied satisfactorily to polypropylenes.

C. APPLICATIONS

Polyethylene film has not been considered competitive to cellophane but has been applied where cellophane was not considered. The packaging market for polyethylene film is shown in Table 51.

Fresh products is the largest category in which cellophane is not a possible competitor. Many of the other categories (meat, poultry, frozen food, and dairy products) are also essentially noncompetitive with cellophane.

A large portion of the cellophane market is in premium grades which cost more but have better properties. Polypropylene has these better properties and offers them at a lower price. Tear resistance is superior and aging characteristics are better. Polypropylene film is a better barrier than many grades of cellophane; moisture resistance is greater.

The physical properties and economics point to polypropylene applications as greater than the cellophane market. Unlike polyethylene, replacement will be at the expense of cellophane. Areas of the polyethylene market where clarity and stiffness are primary considerations will also fall to polypropylene. About 100 million pounds a year of films, such as Pliofilm Vinyl, Saran, etc., are consumed in packaging.

The weakness of cellophane is its moisture sensitivity. Although moistureproofing improves this, cellophane must be kept at narrow moisture limits for optimum properties. Dried products draw the moisture out of cellophane, causing it to

TABLE 51

ESTIMATE OF ANNUAL CONSUMPTION OF FILM IN PACKAGING (MILLIONS OF POUNDS)

Application	Polyethylene 1962 − 1964	Polypropylene 1965	Polypropylene 1970
Baked goods	3.0	80	250
Confectionery	5.0	10	30
Dairy products	6.0	5	30
Drugs	2.0		
Fresh products	75.0	50	200
Frozen foods	7.0	15	80
Garment and shirt bags	35.0		
Meat and poultry	6.0	15	60
Other foods	5.0	35	150
Paper products		15	50
Snacks	2.0		
Textile products	13.0	20	70
Tobacco	1.0	5	30
All others	40.0		
Total	200.0	250	950

embrittle and break in use. Such products are more suitably packaged in polypropylene; for example, flour products, dried milk and eggs, dessert powders, powdered starch, cake mix, biscuit mix, roll mix, rice and oats, and noodles and spaghetti.

Cellophane is limited by its low tear to packages weighing less than 3 lb. Materials packaged in larger units will be better in polypropylene.

Applications for polypropylene film exist in baked goods, fresh products, frozen foods, meat and poultry, confectionery, dairy products, miscellaneous food items, textiles, tobacco, and paper. The baking industry is the largest single market for transparent, flexible packaging materials. Despite growth of polyethylene film in this field, and a minor amount of polypropylene film, 85% is cellophane. Approximately half was used to wrap bread. Out of 8 or 9 million loaves of white bread, 1.5 to 2 million loaves were wrapped in cellophane; the remainder were waxed-paper covered.

A 0.75-mil polypropylene film will function in bread-wrapping equipment as well as 1-mil polyethylene or cellophane. The approximately 75 million pound cellophane market, plus three or four times this amount as waxed paper, could be replaced by polypropylene.

Polypropylene gives (1) longer preservation of freshness, (2) a fresh feel, (3) better resistance to tearing than cellophane or paper, (4) a reclosure by twisting or folding the film, (5) preservation of flavor and retardation of mold, and (6) storage under refrigeration without becoming brittle. Thick sections of molded polypropylene become brittle at low temperatures; thin films retain strength and flexibility at deep-freeze temperatures. Polypropylene film used in wrapping bread does not transfer anything to the bread.

The baked-goods industry offers a market which should take up to about 400 million pounds of transparent film, a large part of which should be polypropylene film.

Fresh products are the largest market, consuming about 75 million pounds a year. The market available for polypropylene for fresh products is given in Table 52.

TABLE 52

Crop	Average annual supply, thousands of pounds	Estimated % packaged, 1965	Estimated potential for polypropylene, thousands of pounds
Apples	3,500	35	4.5
Beans	550	50	2.5
Beets	150	50	0.5
Carrots	1,500	50	4.0
Celery	1,500	40	5.0
Corn	1,500	25	2.5
Grapefruit	1,800	20	1.5
Grapes	900	20	1.5
Lemons	600	20	0.5
Lettuce	3,300	50	10.0

Limes	30	20	0.2
Onions, dry	1,800	35	2.5
Oranges	4,800	20	3.5
Parsnips	60	35	0.1
Peas, green	60	25	0.5
Potatoes, white	18,000	30	7.5
Potatoes, sweet	1,500	30	1.5
Radishes	300	60	1.0
Tangerines	350	20	0.5
Turnips	400	50	0.5
Retail level packaging and items not listed			5.0
Liners			5.0
Estimated total potential for polypropylene			60.3

The frozen-food market today includes about 1000 different items with a yearly volume of 6 or 7 billion packages.

Product	Billions of packages annually
Vegetables and fruits	1.5
Precooked foods and specialties	1.5
Seafood.	0.5
Meat, meat products, and poultry	2.0

Frozen foods are packaged in a paper-board covered with a printed, waxed-paper wrap.   This is attractive but is expensive.   A trend is therefore toward a polyethylene-bag pack, which could be replaced by polypropylene.

The boilable pouch can serve both as package and as container for food preparation. Strength retention at boiling-water temperature has required polyesters. Polypropylene has ample strength at the temperature of boiling water for this application. Polyethylene consumption for frozen foods is 7 or 8 million pounds a year, and polypropylene can hardly be expected to displace polyethylene.

About half of all fresh meats sold are prepackaged, mostly from self-selection meat cases, and the trend is to increase prepackaging.   Cellophane dominates the field by virtue of special coated grades with the proper relationship between moisture control and oxygen transmission.

Poultry packaging consumes transparent films, such as Pliofilm, Saran, and Cry-o-vac. These conform to the contour of the poultry, making an attractive package. Polyethylene, in spite of its lower price, is used only to a small extent. Applications of polyethylene film in sausage, bacon, ham, and processed meats are open to market penetration by poly-propylene.

Almost 3 billion pounds of confectionery products are produced in the United States annually, and about 5 million pounds of polyethylene film packaged it. Polyethylene has almost ideal characteristics for this application. Confectionery depends to a great extent on impulse buying, and even a little haze in the film can be an obstacle.

In dairy products polyethylene film is used for dry powdered milk and liquid cream. Polypropylene should be a competitor because of its superior strength.

Transparent film wrap in the textile field has grown rapidly. Some advantages are (1) protection from soiling, (2) allowing inspection, and (3) appealing appearance.

Polyethylene film is important in packaging blankets, sheets, pillowcases, shirts, sweaters, infants' wear, children's wear, women's gloves, pillows and draperies, yard goods, hosiery, napkins, tablecloths, pajamas, underwear, and household furnishings. Vinyl film holds much of the sheet and pillowcase packaging market.

Cellophane is entrenched in the cigarette market: 40 million pounds are used yearly to package cigarettes. Polypropylene film has characteristics to take over portions of this market. Several million pounds of cellophane are also used as cigar wrappers and as overwrappers on five-packs and larger boxes.

For pouch material for pipe tobacco, laminates of paper and polyethylene are used. Polypropylene film should be strong enough to use without the paper reinforcement, and also moisture-resistant enough to make other barrier materials, such as aluminum foil, unnecessary. Polypropylene-film applications are small in volume and must be considered experimental.

Shrink packaging has grown rapidly. Nine shrink films compete, as shown in Table 53. These films offer tensile strengths from 1600 to 25,000 psi, water-vapor transmissions of 0.2 to 14.0 g/100 in.^{2} per 24 hr/mil, and oxygen permeabilities of 20 to 10,000 cc/m^2 per 24 hr/mil.

The heat shrink characteristics are built into the film during manufacture by (a) stretching under controlled temperatures and tensions and (b) locking the film in stretched condition by cooling. Stored shrink energy is later released by heating to soften the polymer or melt the crystals, allowing the film to pull back toward its original unstretched condition. Some of the high-shrink films, such as the cross-linked

TABLE 53

SHRINK FILMS

Name of film material	Tensile strength, psi	Water-vapor permeation in g/100 in.^{2}/ 24 hr through a 1-mil film at 90% relative humidity at 100°F
Polyester (Mylar)	up to 28,000	2.5−3.5
Polyethylene A Film made with a polymer in the density range from 0.910 to 0.925	1600−2000	0.6−0.9
Polyethylene B Film of type A, but crosslinked with gamma radiation	up to 15,000	0.5−0.6
Polyethylene C Film made with a polymer in the density range from 0.926 to 0.945 and crosslinked	up to 20,000	0.3−0.4
Polypropylene	up to 28,000	0.2−0.5
Polystyrene	9000−12,000	4.0−9.0
Polyvinylchloride	up to 18,000	3−15
Polyvinylidene chloride	up to 20,000	0.2−0.8
Rubber−HCl (Pliofilm)	up to 12,000	1.0−1.5

Oxygen permeation in $cm^3/m^2/$ 24 hr through a 1-mil film at room temperature at 1 atm pressure	Maximum film shrinkage, %	Film shrinkage temperature, °F	Heat-sealing temperature, °F
20–40	25–35	150–250	–
10,000	15–20	150–250	300–400
7500	70–80	150–250	350–500
3500	70–80	200–300	350–500
1500	70–80	220–330	350–450
3500	50–60	210–270	250–300
20–100	50–70	150–250	250–350
15–40	50–70	150–250	220–280
3000	40–60	150–250	220–280

TABLE 54

PERMEABILITY OF POLYETHYLENE AND POLYPROPYLENE FILMS

Gas[a]	Vapor[b]	Polypropylene	Polyethylene
Carbon dioxide		0.60	1.44
Oxygen		0.18	0.35
Hydrogen		0.70	0.89
Nitrogen		0.04	0.12
Methane		0.08	0.41
	Water	0.20	0.35
	Ethyl alcohol	0.051	0.24
	Ethyl acetate	3.13	11.80
	Acetone	0.30	2.35
	n-Heptane	131.00	120.00
	Toluene	93.60	180.00
	Carbon tetra-chloride	221.00	230.00

[a] In g/100 in.^{2} in 24 hr through a 1-mil film at 1 atm and 75°F.
[b] In g/100 in.^{2} in 24 hr through a 1-mil film at 100°F and 90% relative vapor pressure.

polyethylenes, when heated to a sufficiently high temperature will shrink back to the original size and shape they had prior to the orientation operation.

The shrink available varies from 15 to 20% for polyethylene (Type 1) to 70 to 80% for the cross-linked polyethylene films. In many cases 5 to 10% of shrink is needed. For a contour fit on an odd-shaped product, shrinkage of 50% or more is desirable.

All films change their properties when shrunk. The changes include (a) increase in thickness proportional to reduction of area, (b) decrease in tensile strength, (c) loss of flexibility and increase in stiffness, (d) increased tear resistance, (e) decreased shock resistance, (f) increased abrasion resistance, (g) decrease of transparency, (h) change of surface tack, (i) loss of tension developed during shrinking, and (j) loss of elasticity. Biaxially oriented films are elastic and can be stretched during packaging or subsequent handling without permanently deforming the package.

Permeability (P) is the product of solubility (s) and diffusion (D); the first, an equilibrium and the second, a rate process. The result of their cooperation can be expressed by

$$P = s \cdot D$$

Table 54 contains permeability data of polyethylene and polypropylene film for five gases, and data for seven vapors.

There are processes to allow dyeing and printing with soluble dyestuffs and resin-bonded pigments.

There are a large number of overwrap market possibilities for polyolefin films and portions of these may fall to quenched, unoriented polypropylene because of its transparency, dimensional stability, sparkle, or toughness. Thus, overwraps for textiles, moisture barriers for tobacco pouches, cigarette packages, chewing gum, and bread wraps represent possibilities. As fabricated bags from film for candy, produce, dried foods, and other hydroscopic products, there are interesting developments and possibilities for polypropylene. Market evaluation is needed as well as large-scale testing in the application of moisture-barrier film for construction, bags for bulk chemicals and protection of dry-cleaning apparel.

There are some potential markets for polypropylene not hitherto open to simple unlaminated films. Because film from the slot die process can be heated to 270°F with only minor shrinkage, there is the possibility of pouch packaging of contents that must be sterilized, such as baby foods, utensils, surgical instruments, and the like. There is good potential in cook-in pouches for fresh meats and many fresh or frozen vegetables, or in the form of low gas-transmission material for smoked meats and oxidation-sensitive foods. This is quite apart from shrinkable packaging for frozen meats and poultry. The rate of the development is a function of the manufacturers of polypropylene film, the converters, and the competitive position of other films from the viewpoint of the consumers. Polypropylene film is definitely a commercial economic material in relatively wide use at the present, and with large possibilities for the future.

References

1. Anonymous, Eastman Kodak Company, Plastics Division, Technical Report, Monofilament Extrusion with Tenite Polypropylene, No. TR-6, 1961.
2. Anonymous, Eastman Kodak Company, Plastics Division, Technical Report, Effect of Polypropylene Flow Rate on Monofilament Properties, No. TR-11.
3. Anonymous, Hercules Powder Company, Fiber Development Department, Electrical Behavior of Herculon Polypropylene Fiber, Bulletin FD-10, November 26, 1962.
4. Anonymous, Hercules Powder Company, Polymers Department, Pro-fax, Information on Hercules Polypropylene, Bulletin, 1959.
5. Bernhardt, E. C., Processing of Thermoplastic Material (New York, Reinhold Publishing Corporation, 1960).
6. Billmeyer, F. W., Textbook of Polymer Chemistry (New York, Interscience Publishers, Inc., 1959).
7. Cappuccio, Coen, A., Bertinotti, and Conti, W., "Fibers from Isotactic Polypropylene," Chim. Ind. 44:463–73 (1962).
8. Carroll-Porczynski, C. Z., "Manual of Man-Made Fibers" (New York, Astex Publishing Co., 1953).
9. Coen, A. and Conti, W., "Physical and Chemical Properties of Polypropylene Monofilaments," Materie Plastiche 26:723–30 (1960).
10. Cohen, E. Z., "Meraklon Polypropylene Fibers," Am. Dyestuff Reptr. 51:596–600 (1962).
11. Compostella, M., Coen, A., and Bertinotti, "Fibers and Films of Isotactic Polypropylene," Angew. Chem. 74:618–24 (1962).
12. Dorset, M., "Discussion of New Fibers and Their Properties," Textile Mfr. 87:503–507 (1961).
13. Erlich, V., Mod. Textiles Mag. 39 (No. 11):59–67 (1958).
14. Fisher, E. G., Extrusion of Plastics (New York, Interscience Publishers, Inc., 1958).
15. Floyd, D. E., Polyamide Resins (New York, Reinhold Publishing Corporation, 1958).
16. Goppel, J. M., "Chemistry and Properties of Polypropylene," Brit. Plastics No. 5:207–12 (1959).
17. Hall, A. J., "Polyethylene and Polypropylene — The Versatile Polyolefin Fibers," Fibres Plastics 22:5–9 (1961).
18. Hall, I. H., "The Effect of Temperature and Strain Rate on the Stress–Strain Curve of Oriented Isotactic Polypropylene," J. Polymer Sci.: 54:505–22 (1961).
19. Hamburger, W. J., "Mechanics of Abrasion of Textile Materials," Textile Res. J. 15:169 (1945).
20. Harris, Milton, Handbook of Textile Fibers (Washington, D. C., Harn's Research Laboratories, Inc., 1954).
21. Henstead, W., "Polyolefines in Textiles," J. Textile Inst. 52:158–68 (1961).
22. Hooper, G. S., "Polypropylene Textile Fibers," Textile Res. J. 32:529–39 (1962).

23. Kaswell, E. R., "Textile Fibers, Yarns, and Fabrics," (New York, Reinhold Publishing Co., 1953).
24. Kaswell, E. R., "Wellington Sears Handbook of Industrial Textiles," (New York, Publications Dept., Wellington Sears Co., 1963).
25. Kresser, T. A., Polypropylene, (New York, Reinhold Publishing Co., 1960).
26. Laible, R. C. and Morgan, H. M., "Viscoelastic Behavior of Isotactic Polypropylene Fibers," J. Appl. Polymer Sci. 6:269—77 (1962).
27. Mahn and Vermillion, "Polypropylene Monofilament Extrusion," Plastics Technol. 7: 23—25 (1961).
28. Mauersberger, "American Handbook of Synthetic Textiles" (New York, John Wiley and Sons, Inc., 1962).
29. Natta, G., "A New Italian Demonstration in the Textile Industry: The Polypropylene Fibre," Chim. Ind. 41 (No. 7):647—652 (1959).
30. Press, J. J., Man-Made Textile Encyclopedia (New York, Interscience Publishers, Inc., 1959).
31. Roberts, J. F. L., "Isotactic Polypropylene Fibers," Mineralogical Chemistry; W. H. Bennett, Rept. Progr. Appl. Chem. 45:367—72 (1960).
32. Sheehan, W. C. and Cole, T. B., "Production of Supertenacity Polypropylene Filaments," Southern Research Institute, Birmingham, Alabama (to be published).
33. Thompson, A. B., "Polypropylene Fibers," Man-Made Textiles: 38 (No. 445):35—38 (1961).
34. Thompson, A. B., "Fibers from Polypropylene," J. Royal Inst. Chem. 85:293—300 (1961).
35. Wyckoff, H. W., J. Polymer Sci. 62:83 (1962).

Definitions and Fiber Trademarks

Spin-draw ratio
>   Ratio of take-up speed to extrusion speed in the spinning process.

Draw ratio
>   Ratio of the length of the drawn filament to the undrawn filament; draw ratio may also be expressed as the ratio of the denier of the undrawn fiber to the denier of the drawn fiber.

Total draw ratio
>   Product of the spin-draw ratio and the draw ratio.

Dacron
>   Polyester filament; trademark of Du Pont de Nemours and Company.

Orlon
>   Acrylic fiber; trademark of Du Pont de Nemours and Company.

Herculon
>   Polypropylene filament; trademark of the Hercules Powder Company.

Terylene
>   Polyester fiber; trademark of the Imperial Chemical Industries, England.

Ulstron
>   Polypropylene fiber; trademark of the Imperial Chemical Industries, England.

Tenite
>   Polypropylene; trademark of the Eastman Kodak Company.

Fortisan
> High-tenacity saponified acetate filament; trademark of the
> Celanese Fibers Company.

Profax
> Polypropylene; trademark of the Hercules Powder Company.

INDEX